Migration, Land and Livelihoods

This book critically and succinctly examines recent changes in land ownership, mobility and livelihoods in various Pacific island states, from East Timor to the Solomon Islands, where climate change, environmental change (including hazards of various origins), population growth and urbanization have contributed to new tensions and discords and resulted in complex structures of migration and re-settlement. This has brought new and varied experiences of income and livelihood generation, and consequent reinterpretations of 'modernity' and 'tradition'. In a series of detailed case studies this book traces various responses to such socio-economic changes both in how they are locally envisaged, as pressures on land have intensified, urban informal settlements and livelihoods have expanded and per-ceptions of identity and property rights have changed, and in national development policy responses. It offers valuable reflections on the complex balance between continuity and change, the tensions between social and economic development, the will to develop and the management of dissent and difference.

This book was originally published as a special issue of *Australian Geographer*.

George N. Curry is Professor of Geography in the Faculty of the Built Environment at Curtin University, Perth.

Gina Koczberski is a Senior Research Fellow in the Faculty of the Built Environment at Curtin University, Perth.

John Connell is Professor of Geography in the School of Geosciences, University of Sydney.

Migration, Land and Livelihoods

Creating Alternative Modernities in the Pacific

Edited by
George N. Curry, Gina Koczberski and John Connell

Routledge
Taylor & Francis Group

LONDON AND NEW YORK

First published 2015
by Routledge
2 Park Square, Milton Park, Abingdon, Oxon, OX14 4RN, UK

and by Routledge
711 Third Avenue, New York, NY 10017, USA

Routledge is an imprint of the Taylor & Francis Group, an informa business

British Library Cataloguing in Publication Data
A catalogue record for this book is available from the British Library

ISBN 13: 978-1-138-80398-5

Typeset in Plantin
by RefineCatch Limited, Bungay, Suffolk

Publisher's Note
The publisher accepts responsibility for any inconsistencies that may have
arisen during the conversion of this book from journal articles to book chapters,
namely the possible inclusion of journal terminology.

Disclaimer
Every effort has been made to contact copyright holders for their permission to
reprint material in this book. The publishers would be grateful to hear from any
copyright holder who is not here acknowledged and will undertake to rectify
any errors or omissions in future editions of this book.

Contents

Citation Information

Please direct any queries you may have about the citations to
clsuk.permissions@cengage.com

Citation Information

The following chapters were originally published in *Australian Geographer*, volume 43, issue 2 (June 2012). When citing this material, please use the original page numbering for each article, as follows:

Chapter 1
[...]
Australian Geographer, volume 43, issue 2 June 2012, pp. 115–129

Chapter 2
[...]
Australian Geographer, volume 43, issue 2 June 2012, pp. 131–143

[...]

[Please direct any queries you may have about the citations to clsuk.permissions@cengage.com]

Introduction: enacting modernity in the Pacific?

GEORGE N. CURRY, GINA KOCZBERSKI & JOHN CONNELL,
Curtin University, Australia; Curtin University, Australia; University of Sydney, Australia

Rural and urban communities in the culturally and economically diverse Pacific region are experiencing significant economic, environmental and social change as they grapple with the challenges and opportunities of globalisation, modernity and environmental change. Earlier views of an inevitable and linear transformation of indigenous societies as they became incorporated through globalisation of economies, societies and cultures are now giving way to more nuanced under-standings of the complexities of change and the importance of local agency in shaping the direction and pace of change. With greater attention to local-level factors and the role of agency, the notion of pre-determined outcomes is being questioned by the recognition that change is neither unidirectional nor, indeed, a scripted transition from a pre-capitalist economy to a capitalist one or from tradition to modernity; rather, the outcome is messier and the indeterminacy of the encounter with various notions of capitalism means that a diversity of local outcomes is to be expected.

A challenge is how to conceptualise and understand these processes at various scales, and the relationships across scales, and in quite different places. The geographer's task, then, is to disentangle these multiple processes and logics to understand how these diverse modernities are created and what they mean for, and how they are valued by, local communities experiencing and contributing to these changes. The papers in this issue, which emerged mainly from a special session on 'Migration, Land and Livelihoods' at the 2009 annual IAG conference, all adopt a local-level approach to explore processes of change in some of the countries to Australia's north, although all are cognisant of macro-level processes and how they interact with local-level factors to shape social and economic outcomes at the local level. Whilst migration and access to land are common threads to this collection, the papers make a broader contribution to ideas and concepts in development, particularly to recent ideas on the enduring influence of indigenous economic and social forms and their role in creating alternative modernities.

Land

Land in most agriculturally-based Pacific societies is more than an economic asset. Land holds a prominent position in providing sustenance, cultural and spiritual

1

beliefs, in social and ritual activities, social organisation and in creating an individual and group's sense of social identity and belonging. Eighty years ago on the east coast of New Caledonia, the distinguished ethnographer/missionary Leenhardt noted that society was 'written on the ground'. Melanesians were so closely tied to their land that it defined who they were—land was crucial to the past, essential for the present and critical for the future. Membership in landholding groups not only continues to provide access to land but it also defines who people are, while ties between land and identity forge an eternal bond of belonging to place. Land was associated with the ancestors, and 'rocks, plants and the human body originate in similar structures' (Leenhardt 1947/79, p. 61), and place, language and time were tied together. Indeed, the past may be understood spatially as much as temporally (Kahn 2011). As Sillitoe points out for the Wola of the Highlands of Papua New Guinea (PNG):

> In a sense Wola belong to the land as much as it belongs to them ... They have a deep affection for their land, talk warmly about it—how their father or ancestor cultivated where they currently have gardens—and relate events witnessed in the vicinity, such as battles, ceremonies and so on. Land has an everlasting aspect; it is their connection with the past and the future ... All kin, including dead ancestors and unborn descendants, depend on the same land. (Sillitoe 1999, pp. 348–9)

This intense attachment to land and its centrality to social identity underpins the widely held view amongst customary landowners that land is inalienable. Even at massively transformed mine sites landowners rarely migrate away from the project and its disturbances. To be landless is an unimaginable and inconceivable misfortune. All of this helps to explain why in many areas of the Pacific the notion that land can be alienated permanently is unacceptable to customary landowners and is the root cause of much conflict across the island arc from Timor eastwards through the Pacific Islands.

Post-contact

Despite the centrality of land in culture and society, customary land tenures are also dynamic and have been modified considerably in response to changing circumstances and new opportunities associated with colonialism, capitalism and the emergence of market economies (Crocombe 1971; Ward 1995; Ward & Kingdon 1995; Strathern & Stewart 1998; Martin 2007). Whilst large tracts of customary land were alienated—mainly for plantations in the pre-independence period—much has remained under customary ownership or has been returned to customary ownership since independence. However, land has increasingly been placed under new stresses as a result of population pressures and new demands on land for development. As the papers in this collection reveal, three key drivers of change include large-scale resource development pressures (mining, logging and commodity crop production), urbanisation and internal migration, while environmental hazards have also precipitated and necessitated new approaches to land tenure.

Across the region land tenure is being modified at a range of scales from the 'land-grabbing' efforts of mainly international companies seeking access to large tracts of land, often for logging, mining and commercial crop production but also for urbanisation, or even tourism (Filer 2011; Slatter 2006). Areas of large-scale commodity crop production and mine sites are becoming hotspots for conflict as these projects of modernity and development are associated with significant in-migration from poor, peripheral regions, thereby placing additional demands on customary land. The scale and nature of this in-migration often irreversibly restructures local relationships so that local communities may feel dispossessed, marginalised or displaced not just from the project but also from the sheer number of people arriving on their lands (Banks 2009; Bainton 2010).

Throughout the Pacific, people are leading increasingly mobile lives as they seek new livelihood opportunities. Whilst much of this migration is rural-to-urban, a sizeable proportion is rural-to-rural, particularly to agricultural and resource frontier zones where communal tenure is the dominant form of land tenure. In pursuit of livelihoods, land-poor migrants are entering into a broad range of formal and informal arrangements with customary landowners to gain short- and long-term access to land, thereby creating burgeoning informal land markets and changes to customary land tenure regimes. These movements, some of which are now multi-generational, mean that village communities are becoming more dispersed, leadership more fragmented, and 'real' landowners multiplying as marriage networks widen spatially and culturally and migrants make claims on the land of their hosts. People's intimate attachments to land are therefore weakening, allowing land to be viewed more as a commodity. As none other than Francis Fukuyama has pointed out,

> The fact that there are multiple classes of claimants to a particular parcel of land, no strong tradition of delegated authority, and no statute of limitations with regard to customary claims, means that it is extremely difficult ... to come up with schemes by which landowners can pool resources to convert customary land into modern alienable property. (2008, p. 21)

But that supposes that landowners wish to be entirely engaged in the capitalist system. In most cases they do not.

The production of commodity crops not only places more pressure on land by removing land from subsistence production, thereby intensifying pressure on the subsistence base, but also requires significant changes to land tenure as land is converted from annual cropping over a few seasons to perennial export cash cropping. As more land is devoted to commercial agriculture and comes under semi-permanent commercial production, usufruct rights remain vested in the same families or individuals for very long periods. This long-term alienation of land for commercial purposes is leading people to claim exclusive rights of access to, and inheritance of, these resources. The outcome is that communal tenure is giving way to individual tenure with a consequent loss of flexibility in the allocation of land rights. However, as detailed studies have increasingly shown, awareness of the problems attached to the loss of flexibility has discouraged landowners from moving in that direction, even where pressures in the vicinities of mines would seem to offer tremendous opportunities for landowners to capitalise on economic rents.

Traditional land tenure can accommodate change

Whilst migration has been a focus of research amongst geographers, less attention has been paid to understanding the local-level processes and mechanisms by which migrants without land use rights gain access to customary land and other resources to construct livelihoods in their new homes. Although the flexibility and adaptability of customary land tenure practices in PNG and the Pacific that enabled 'outsiders' to cultivate temporary food gardens on another's customary land has been noted (e.g. Crocombe & Hide 1971; Jorgensen 2007), the gifting or 'selling' of land for the cultivation of perennial export tree crops or for long-term residence was much less common, especially so when the 'outsiders' were from another ethnic group and lacked marital or trading ties with the customary landowning group (Curry 1997). Thus recent changes associated with resource developments where large numbers of migrants are settling and gaining relatively long-term access to land—over several generations in some cases—is leading to complex land tenure arrangements. These transactions involve difficult choices for both migrants and landowners, as reflected in this set of papers. At mine sites particularly this has led to rapidly evolving and complex contexts where the claims of multiple landowners challenge the abilities of local and national governments and mining companies to regulate or cope with the speed of change and the diversity of claims. Legal recognition of the complex variety of customary forms of landowner-ship has ensured that companies operating in the Pacific are drawn into intense and often continuing engagement with local landowning groups over compensation, royalty payments, the services and employment that mines particularly deliver and their environmental and social impacts. Indeed, the relationships between land, people and mine sites have extended the issues that surround the limits and boundaries of corporate social responsibility (Banks 2006). In 1988 the Bougainville copper mine in PNG closed down after unresolved tension over such issues spilled over into sustained violence (Filer 1990; Connell 1991).[1] The papers explore gradual or evolutionary processes of change as well as rapid and sometimes conflicted change—urbanisation (Numbasa & Koczberski this issue), rural-to-rural migration (Allen this issue; Koczberski *et al.* this issue), and the migration of whole groups from hazardous environments (Connell this issue); and, in East Timor, the legacy of late colonial pressure with forced resettlement (Thu this issue). Some-times these new land tenure arrangements entail the superimposition of alien ideologies and practices which are usually in some way superficial, while others such as Western models of land tenure promoted by national governments, aid organisations and consultants remain 'foreign flowers' (Larmour 2005) that never do fit.

What is clear from the papers in this issue is that for most customary landowners, land rights granted to private companies, migrants or the state are rarely seen as permanent and exclusive. Customary landowners' ongoing sense of 'ownership' of their 'alienated' land adds a layer of uncertainty to land transactions. Notions of partial alienation of customary land raise new questions of equity and uneven development and the nature of identity, tied as it is to land, for both landowners and migrants (Koczberski & Curry 2004; Allen this issue). These issues become even more complicated by the sudden arrival of migrants after hazard events (Connell this issue), or the short-term nature of many mining operations, which can induce landowners to appropriate mining benefits disproportionally, thereby marginalising migrants (Banks 2008).

Embedded in social relationships

The papers in this issue fit into a growing body of literature from the Pacific region that shows how indigenous economic logics shape contemporary economic practices and values in such a way as to reconfigure people's relationship with global capitalism to give it cultural meaning (Gregory 1982; LiPuma 1999, 2000; Curry 2003; Jolly 2005; Sahlins 2005; Connell 2007a, b; Cahn 2008; Curnow 2008; McGregor 2009; Thornton *et al.* 2010; Patterson & Macintyre 2011; Curry & Koczberski 2012). Each case study highlights how customary land tenure regimes and other established social structures have a capacity to accommodate change while remaining grounded in local social and cultural institutions, making generalisations difficult (e.g. Curry 2003; McDougall 2005; Connell 2007b; Curry & Koczberski 2009). Many Pacific people have sought to combine a greater degree of cultural continuity with the impossibility of denying the necessity for economic development and so retreating from modernity. Excursions into capitalism, in whatever variant, have therefore increasingly been seen as cautious essays in economic hybridity that did not, however, contest or reject the renewed expansion of capitalism (Gegeo 1998; Yang 2000; Connell 2007a) but simply sought more successful and more culturally sensitive forms of accommodation to it. At every site of change, complex, culturally encoded changes are played out. Terms such as entanglement, hybridisation, syncretism, multiple modernities, alternative modernities, and cultural appropriation have been used to refer to this complex weaving together of old and new, tradition and modernity, indigenous and foreign and non-market and market (LiPuma 2000; Evans 2001; Horan 2002; James 2002; Curry 2003; Wardlow 2006; Curry & Koczberski 2009; Besnier 2011). As Sahlins (2000, p. 9) notes: 'in all change there is continuity' and the outcome is a diverse array of distinctively different place-based economic and social forms.

This process of hybridisation or blending of indigenous and introduced forms of economy and social values can be thought of, in a Polanyian sense, as the embedding in social relationships of introduced economic forms and values (Polanyi 1944; Polanyi *et al.* 1957; see Curry & Koczberski 2012 for a fuller discussion). Whilst land transactions seemingly entail the commodification of land, access rights are gained through people negotiating socio-economic and political relations that are grounded in place-based social practices and values that draw on an indigenous morality and non-market relationships. In negotiations with customary landowners for access to land, 'outsiders', whether companies or migrants, must often straddle both the market economy and the indigenous non-market economy to first initiate and then maintain relationships with their 'host' landowners (Allen this issue; Koczberski *et al.* this issue; Numbasa & Koczberski this issue; Thu this issue). By embedding land use rights and practices in social relationships with customary landowners, outsiders are able to legitimise their land claims through an indigenous morality that allows them long-term access rights to the land of their hosts. As social networks evolve outsiders may be re-categorised from strangers, even enemies, to guests. Yet, despite the diverse social relationships migrants pursue to gain access rights, there remains a great deal of fluidity and uncertainty in these relational economies of land, and in most contexts, whether at mine sites, settlement schemes or in urban centres, migrants invariably remain vulnerable to the decisions and actions of landowners.

The persistence in modified form of relational economies of land tenure in capital-intensive mining and plantation zones may seem paradoxical. As discussed above, these new forms of tenure arrangements do not involve the displacement of indigenous land tenure by Western land tenure systems, despite sustained efforts by colonial powers and by major multilateral organisations since then, and nor are they simply reproducing traditional forms. Rather, they involve a reworking of customary land tenure which, while still compatible with long-standing principles of indigenous land tenure, has clearly been modified to meet the new requirements of the cash economy and resource development needs. By acknowledging the role of social relationships in accessing resources, the papers in this issue bring a different perspective to the study of global and local interactions by according greater recognition to the role of place-based, local-level factors in shaping trajectories of change and the governance of resource rights.

What is also clear from this set of papers is the capacity of customary landowners and outsiders—companies and migrants—to develop new forms of tenure systems that are able to move beyond the limitations of the state in land matters. In many ways, the state lags behind what is happening on the ground. In PNG, for instance, the weak and diminishing authority of the state in land matters and the growing assertiveness of landowners in their relationships with the state and capital is illustrated by the rising numbers of compensation claims and other landowner demands on government and developers, and the strong bargaining position of landowning groups in resource development negotiations on their land (Filer 1997; Standish 2001; Banks 2008; Bainton 2009). As these papers show, there is much pressure on land in resource development sites, and in the absence of effective state management of land transactions people on the ground negotiate their own ways of resolving these issues as best they can.

Policy implications

The papers in this issue illustrate the rising development pressures on customary land, the emergence of tensions and social conflict between customary landowners and migrants associated with growing levels of mobility characterising contemporary life in the region. Within this changing social and economic milieu, land has become a major planning problem for urban development, mining and agricultural development. For many external observers the complexities of customary land tenure are a major brake on development and a substantial constraint on leasing or selling land and thus establishing a viable private sector necessary for economic development (Gosarevski *et al.* 2004). Many countries have consequently attempted to pursue land reform programs to convert customary land to individual property rights in the form of freehold or individual leasehold title. Yet in societies where land is much more than an economic asset— where society is 'written on the ground'—land ownership is complex and not easily amenable to translation into Western codes and conventions. In Ranongga, Solomon Islands, for example, for 50 years outside agencies have consistently called for the clarification of property rights as the necessary starting point for any form of modern economic development. Local residents are eager to have their rights recognised by outsiders and by other islanders, but transforming complex, cross-cutting localised relationships into abstract rights that are commensurable, predictable and knowable to outsiders raises major political and ethical dilemmas for Ranongga leaders. Claiming exclusive rights to land for oneself or one's group would negate long-standing elements of reciprocity and the relative ease of incorporation of outsiders, and would effectively alienate those others who are essential for the proper functioning of the local polity. Westernisation of land tenure thus threatens the tenuous achievement of unity that Ranonggans see as the prerequisite to peace, prosperity and, as they understand it, proper development (McDougall 2005). In other words, there are real advantages for local development when land tenure remains flexible and even subject to competing claims rather than being finalised and fossilised through the imposition of Western models of individual and alienable land rights so as to become a source of overt contention.

As the papers in this issue show, landowners are not seeking a radical change to customary tenure principles to enable them to free up their land for development projects or to capitalise on the income opportunities of migrants' demand for land. Landowners want to retain 'ownership' and control of their land and wish to see migrants' access rights socially embedded within a relational economy (e.g. sharing of wealth and maintaining indigenous exchange relationships). Acknowledging such sentiments and values is critical for developing culturally appropriate policies to mobilise customary land for development. Moreover, given the failure of land tenure reform that sought to impose Western notions of land tenure (individual and alienable rights), it is now recognised that adaptation, not replacement, of customary tenures is the most appropriate way to resolve land issues for future developments (Fingleton 2005; Koczberski et al. this issue). New local land usage agreements can be developed that draw on customary principles of land tenure—the relational dimensions of the economy—while providing improved tenure security for migrants cultivating cash crops. These new agreements address the need for a relational dimension to land transactions which gives them greater validity in the eyes of customary landowners than previous attempts to commodify land transactions which inevitably led to conflicts and sometimes the eviction of settlers (see also Curry & Koczberski 2009; Koczberski et al. 2009). However, a significant task in many places is for land tenure agreements to become flexible enough to enable and accommodate the settlement of those displaced through no fault of their own, or those who have migrated to urban areas, where land is particularly scarce and deeply contested. Indeed, particular challenges in rapidly growing capital cities have meant complex and diverse processes over many decades to resolve land tenure (e.g. Goddard 2010; Connell 2011) but which often stubbornly remain unresolved. It is here that the tensions of modernity are greatest.

Concluding comments

The papers in this collection deal with what could be labelled frontier zones where global capital in the form of mining, plantation development and associated programs of resettlement interact with local societies, cultures and economies. In this encounter, customary landowners strive to maintain an indigenous, though certainly modified, system of land tenure and economy that remains grounded in place-based social practices and values imbued with an indigenous morality. What the papers show is that these new forms of tenure arrangements and socio-economic relationships with outsiders have not involved the transformation (or destruction) of indigenous forms of land tenure and social organisation grounded in relational identities, and nor are they locked in an unchanging tradition. Instead, they reflect a fair degree of agency whereby local place-based practices embedded in indigenous social and cultural frameworks are constantly being reworked in the interaction with macro-level processes to create a range of alternative modernities.

The creation of alternative modernities through the inflection of development highlights how at the local level people strive to maintain their identity in the context of change, and desire to forge a modernity compatible with their own registers of value. Modernity embraces both opportunities to escape the constraints of kin and community and desires to 'remake and reawaken the autonomy of that community' (LiPuma 2000, p. 10). An emphasis on locality and on globalisation did not bind together antithetical phenomena. Interplay between the local and the global is at the heart of the battle for a compatible modernity, and the desire to determine destiny. Moreover, an insistence on hybridity, evident in the case studies both here and elsewhere, is not only significant within the relatively small-scale communities discussed here but is central to much wider ongoing state building in what remain relatively new nations (Brown & Gusmao 2009; Richmond 2011). In this way, much of what is valued in local life remains: a sense of community in association with a particular tract of land; shared beliefs and values; a rough equality of material conditions; reciprocity and some degree of community control over the means of production. Such values and virtues have enabled localised forms of autonomy and self-reliance that are not readily transferable into a more globalised world (McDougall 2005), just as many Western values and institutions remain fragile 'foreign flowers' (Larmour 2005).

Whilst globalisation and modernity have resulted in considerable social and economic change, there remain significant elements of continuity as people have seized opportunities to expand their participation in development while at the same time striving to maintain place-based cultural beliefs, ideologies and moral domains that shape everyday decisions and practices, whether they be landowners or poor rural migrants seeking a better life. Yet, in a region where cultures remain vibrant what has become now almost a familiar cultural turn in other parts of the world (Schech & Haggis 2000; Radcliffe 2005, 2006) can still be remarkably absent. While Radcliffe and Laurie argue that 'increasingly development looks to culture as resource and as a significant variable explaining the success of development interventions' (2006, p. 231), within the island Pacific there is little sign that culture, in whatever form, is seen as a resource but much more that it is seen as a brake on hopeful structures of development. However, as evidenced by the papers in this issue, and elsewhere, culture needs to be viewed not as a threat to development but as a core element.

NOTE

[1]   This special issue was intended to include a valuable paper on the relationships between migrants, landowners, governments and the company at one particular Pacific mine site, which would not only have given a better balance to the collection but emphasised the challenges to policy formation in rapidly changing circumstances. Unfortunately at the very last minute the paper was withdrawn because the mining company involved perceived these relationships to be too sensitive to be even reported and sensibly discussed. While the editors of this issue and the author of the paper regard it as a classic form of corporate myopia and paranoia it well reflects the acute sensitivities, frictions and tensions that mark sites of rapid change.

REFERENCES

BAINTON, N. (2009) 'Keeping the network out of view: mining, distinctions and exclusion in Melanesia', *Oceania* 79(1), pp. 18–33.

BAINTON, N. (2010) *The Lihir destiny: cultural responses to mining in Melanesia*, ANU E Press, Canberra.

BANKS, G. (2006) 'Mining, social change and corporate social responsibility: drawing lines in the Papua New Guinea mud', in Firth, S. (ed.) *Globalisation and governance in the Pacific Islands*, ANU E Press, Canberra, pp. 259–74.

BANKS, G. (2008) 'Understanding "resource" conflicts in Papua New Guinea', *Asia Pacific Viewpoint* 49(1), pp. 23–4.

BANKS, G. (2009) 'Activities of TNCs in extractive industries in Asia and the Pacific: implications for development', *Transnational Corporations* 18(1), pp. 43–60.

BESNIER, N. (2011) *On the edge of the global: modern anxieties in a Pacific Island nation*, Stanford University Press, Stanford.

BROWN, A. & GUSMAO, A. (2009) 'Peacebuilding and political hybridity in East Timor', *Peace Review* 21(1), pp. 61–9.

CAHN, M. (2008) 'Indigenous entrepreneurship, culture and micro-enterprise in the Pacific Islands: case studies from Samoa', *Entrepreneurship and Regional Development* 20(1), pp. 1–18.

CONNELL, J. (1991) 'Compensation and conflict: the Bougainville Copper Mine, Papua New Guinea', in Connell, J. & Howitt, R. (eds) *Mining and Indigenous peoples in Australasia*, Sydney University Press, Sydney, pp. 55–76.

CONNELL, J. (2007a) 'Holding on to modernity? Siwai, Bougainville, Papua New Guinea', in Connell, J. & Waddell, E. (eds) *Environment, development and change in Rural Asia-Pacific*, Routledge, London, pp. 127–46.

CONNELL, J. (2007b) 'Islands, idylls and the detours of development', *Singapore Journal of Tropical Geography* 28(2), pp. 116–35.

CONNELL, J. (2011) 'Elephants in the Pacific? Pacific urbanisation and its discontents', *Asia Pacific Viewpoint* 52(2), pp. 121–35.

CROCOMBE, R. (ed.) (1971) *Land tenure in the Pacific*, Oxford University Press, Melbourne.

CROCOMBE, R. & HIDE, R. (1971) 'New Guinea: unity in diversity', in Crocombe, R. (ed.) *Land tenure in the Pacific*, Oxford University Press, Melbourne, pp. 292–333.

CURNOW, J. (2008) 'Making a living on Flores, Indonesia: why understanding surplus distribution is crucial to economic development', *Asia Pacific Viewpoint* 49(3), pp. 370–80.

CURRY, G.N. (1997) 'Warfare, social organisation and resource access amongst the Wosera Abelam of Papua New Guinea', *Oceania* 67(3), pp. 194–217.

CURRY, G.N. (2003) 'Moving beyond postdevelopment: facilitating indigenous alternatives for "development"', *Economic Geography* 79, pp. 405–23.

CURRY, G.N. & KOCZBERSKI, G. (2009) 'Finding common ground: relational concepts of land tenure and economy in the oil palm frontier of Papua New Guinea', *Geographical Journal* 175(2), pp. 98–111.

CURRY, G.N. & KOCZBERSKI, G. (2012) 'Relational economies, social embeddedness and valuing labour in agrarian change: an example from the developing world', *Geographical Research* doi: 10.1111/j.1745-5871.2011.00733.x

EVANS, M. (2001) *Persistence of the gift: Tongan tradition in transnational context*, Wilfred Laurier University Press, Waterloo.

FILER, C. (1990) The Bougainville rebellion, the mining industry and the process of social disintegration in Papua New Guinea, *Canberra Anthropology* 13(1), pp. 1–39.

FILER, C. (1997) 'Compensation, rent and power in Papua New Guinea', in Toft, S. (ed.) *Compensation for resource development in Papua New Guinea*, Law Reform Commission of Papua New Guinea and Resource Management in Asia and the Pacific, Research School of Pacific and Asian Studies, Australian National University, Canberra, pp. 156–90.

FILER, C. (2011) *The new land grab in Papua New Guinea*, ANU State Society and Governance Discussion Paper 2011/2, Canberra.

FINGLETON, J. (2005) 'Conclusion', in Fingleton, J. (ed.) *Privatising land in the Pacific: a defence of customary tenures*, Discussion Paper No. 80, Australia Institute, Canberra, pp. 34–7.

FUKUYAMA, F. (2008) 'State building in Solomon Islands', *Pacific Economic Bulletin* 23(3), pp. 18–34.

GEGEO, D. (1998) 'Indigenous knowledge and empowerment: rural development examined from within', *The Contemporary Pacific* 10(2), pp. 289–316.

GODDARD, M. (2010) 'Heat and history: Moresby and the Motu-Koita', in Goddard, M. (ed.) *Villagers and the city: Melanesian experiences of Port Moresby, Papua New Guinea*, Sean Kingston, Wantage, pp. 19–46.

GOSAREVSKI, S., HUGHES, H. & WINDYBANK, S. (2004) 'Is Papua New Guinea viable with customary land ownership?', *Pacific Economic Bulletin* 19(3), pp. 133–6.

GREGORY, C.A. (1982) *Gifts and commodities*, Academic Press, New York and London.

HORAN, J.C. (2002) 'Indigenous wealth and development: micro-credit schemes in Tonga', *Asia Pacific Viewpoint* 43(2), pp. 205–21.

JAMES, K.E. (2002) 'Disentangling the "grass roots" in Tonga: "traditional enterprise" and autonomy in the moral and market economy', *Asia Pacific Viewpoint* 43(3), pp. 269–92.

JOLLY, M. (2005) 'Beyond the horizon? Nationalisms, feminisms, and globalisation in the Pacific', *Ethnohistory* 52(1), pp. 137–66.

JORGENSEN, D. (2007) 'Changing minds: hysteria and the history of spirit mediumship in Telefolmin', in Barker, J. (ed.) *The anthropology of morality in Melanesia and beyond*, Ashgate, Aldershot, pp. 113–30.

KAHN, M. (2011) *Tahiti beyond the postcard: power, place and everyday life*, University of Washington Press, Seattle.

KOCZBERSKI, G. & CURRY, G.N. (2004) 'Divided communities and contested landscapes: mobility, development and shifting identities in migrant destination sites in Papua New Guinea', *Asia Pacific Viewpoint* 45(3), pp. 357–73.

KOCZBERSKI, G., CURRY, G.N. & IMBUN, B. (2009) 'Property rights for social inclusion: migrant strategies for securing land and livelihoods in Papua New Guinea', *Asia Pacific Viewpoint* 50(1), pp. 29–42.

LARMOUR, P. (2005) *Foreign flowers, institutional transfer and good governance in the Pacific Islands*, University of Hawai'i Press, Honolulu.

LEENHARDT, M. (1947/79) *Do Kamo: person and myth in the Melanesian World*, University of Chicago Press, Chicago (Gallimard, Paris, 1947).

LIPUMA, E. (1999) 'The meaning of money in the age of modernity', in Akin, D. & Robbins, J. (eds) *Money and modernity: state and local currencies in Melanesia*, University of Pittsburgh Press, Pittsburgh, pp. 192–231.

LIPUMA, E. (2000) *Encompassing others: the magic of modernity in Melanesia*, University of Michigan Press, Ann Arbor.

MARTIN, K. (2007) 'Land, customary and non-customary, in East New Britain', in Weiner, J.F. & Glaskin, K. (eds) *Customary land tenure and registration in Australia and Papua New Guinea: anthropological perspectives*, Asia-Pacific Environment Monograph 3, ANU E Press, Canberra, pp. 39–56.

McDOUGALL, D. (2005) 'The unintended consequences of clarification: development, disputing, and the dynamics of community in Ranongga, Solomon Islands', *Ethnohistory* 52(1), pp. 81–109.

McGREGOR, A. (2009) 'New possibilities? Shifts in post-development theory and practice', *Geography Compass* 3, pp. 1688–702.

PATTERSON, M. & MACINTYRE, M. (eds) (2011) *Managing modernity in the western Pacific*, University of Queensland Press, St Lucia.

POLANYI, K. (1944) *The great transformation*, Rinehart, New York.

POLANYI, K., ARENSBERG, C. & PEARSON, H. (eds) (1957) *Trade and markets in early empires*, Glencoe Free Press, New York.

RADCLIFFE, S. (2005) 'Development and geography: towards a postcolonial development geography?', *Progress in Human Geography* 29(3), pp. 291–8.

RADCLIFFE, S. (2006) 'Culture in development thinking: geographies, actors and paradigms', in Radcliffe, S. (ed.) *Culture and development in a globalizing world*, Routledge, London, pp. 1–29.

RADCLIFFE, S. & LAURIE, N. (2006) 'Culture and development: taking culture seriously in development for Andean indigenous people', *Environment and Planning D: Society and Space* 24(2), pp. 231–48.

RICHMOND, O. (2011) 'De-romanticising the local, de-mystifying the international: hybridity in Timor Leste and the Solomon Islands', *The Pacific Review* 24(1), pp. 115–36.

SAHLINS, M. (2000) *Culture in practice: selected essays*, Zone Books, New York.

SAHLINS, M. (2005) 'The economics of develop-man in the Pacific', in Robbins, J. & Wardlow, H. (eds) *The making of global and local modernities in Melanesia: humiliation, transformation and the nature of cultural change*, Ashgate, Farnham, pp. 23–43.

SCHECH, S. & HAGGIS, J. (2000) *Culture and development: a critical introduction*, Blackwell, Oxford.

SILLITOE, P. (1999) 'Beating the boundaries: land tenure and identity in the Highlands of Papua New Guinea', *Journal of Anthropological Research* 55(3), pp. 331–60.

SLATTER, C. (2006) *The Con/Dominion of Vanuatu? Paying the price of investment and land liberalisation—a case study of Vanuatu's tourism industry*, Oxfam New Zealand, Auckland.

STANDISH, B. (2001) 'Papua New Guinea in 1999–2000', *Journal of Pacific History* 36(3), pp. 285–98.

STRATHERN, A. & STEWART, P. (1998) 'Shifting places, contested spaces: land and identity politics in the Pacific', *Australian Journal of Anthropology* 9(2), pp. 209–24.

THORNTON, A., KERSLAKE, M.T. & BINNS, T. (2010) 'Alienation and obligation: religion and social change in Samoa', *Asia Pacific Viewpoint* 51(1), pp. 1–16.

WARD, R.G. (1995) 'Land, law and custom: diverging realities in Fiji', in Ward, R.G. & Kingdon, E. (eds) *Land, custom and practice in the South Pacific*, Cambridge University Press, Cambridge, pp. 198–249.

WARD, R.G. & KINGDON, E. (eds) (1995) *Land, custom and practice in the South Pacific*, Cambridge Asia-Pacific Studies, Cambridge University Press, Cambridge.

WARDLOW, H. (2006) *Sexuality and agency in a New Guinea society: wayward women*, University of California Press, Berkeley.

YANG, M.M. (2000) 'Putting global capitalism in its place: economic hybridity, Bataille and ritual expenditure', *Current Anthropology* 41(4), pp. 477–510.

Population Resettlement in the Pacific: lessons from a hazardous history?

JOHN CONNELL, *School of Geosciences, University of Sydney, Australia*

ABSTRACT *For more than a century Pacific Islanders have been resettled away from difficult circumstances arising from population pressures, environmental hazards and political or economic pressures. Past resettlement strategies and practices provide lessons for the future. Resettlement has raised issues of compensation, land tenure, identity, sovereignty, cultural tensions and establishment of livelihoods. Resources have often been too limited to enable adequate infrastructure and access to services. Mismanagement, marginality and local opposition have posed problems. Many involuntary migrants have sought to return home. In post-colonial times resettlement has been unwelcome and rarely inclusive, with poor outcomes for social and economic development. Significant future hazards are likely to require international responses.*

> Each man must have some place, some land which belongs to him, which is his territory. If he does not control any land, he has no roots, status or power. In the extreme case this means he is denied social existence. (Bonnemaison 1984, p. 1)

In 1856 the entire population—almost 200 people—of isolated Pitcairn Island was resettled by the British colonial government on Norfolk Island, over 6000 km to the west: the longest distance of any official resettlement in the Pacific. Already a first resettlement in Tahiti in 1831 had already failed; a subsequent proposed resettlement to Hawaii had never got off the ground. The tiny island of Pitcairn had been deemed unable to support the population's food requirements. The Pitcairners were granted various land rights on Norfolk, an important drawcard for their migration, but despite assurances of adequate land they gained title to no more than a quarter of Norfolk. Several disappointed Pitcairners returned 'home' to their *fenua maitai* (good land) in two separate episodes (Emery 1985). In many respects the fate of the Pitcairners on Norfolk, marked by inadequate access to land, the challenge of scale, nostalgia and the desire to return home proved extraordinary precedents for a much more recent past history of resettlement in the Pacific.

A widespread assumption exists that resettlement may become a more common Pacific phenomenon in the future, if climate change and unanticipated environmental hazards conspire to make island life more difficult. Similar resettlements have already occurred in multiple contexts, but not without problems. Resettlement has also occurred where land has been resumed by governments for large-scale mining or military exercises and installations, with somewhat similar outcomes as populations have also been collectively resettled. While most migration within and beyond islands in the Pacific region has been of households and individuals, a distinctive response to severe hazards (and some extreme human-induced events) has been more or less formal resettlement schemes over both short and long distances, ranging from small groups of people to entire island populations. While the various migration episodes considered here are very different in their causes (human induced and environmental), severity and consequences, and in the willingness of local people to participate, many have involved more or less formal collective resettlement. This paper reviews the experience of formal resettlement in the Pacific Islands in the post-Second World War era, and examines the extent to which this provides lessons for future resettlement policies and needs. It seeks to examine the rationale for and outcome of these distinctive collective migration moves, and the extent to which they provide a precedent for future Pacific migration.

Human pressures

In the post-war years especially a series of significant migration movements followed human-induced changes, notably from what were deemed overcrowded islands in the Gilbert and Ellice Islands (now Kiribati and Tuvalu) initially to the Phoenix Islands and subsequently to less densely populated parts of what were then other British colonial territories in the Pacific: the Solomon Islands and Fiji.

In the 1930s over 720 Gilbertese had been resettled from the Gilbert Islands to the unpopulated but remote Phoenix Islands in response to perceptions of overcrowding in the main island chain. The instigator of the resettlement, Harry Maude, argued that the optimum population density of each island had been reached by the year 1840 (Maude 1968, p. 319), while in 1865 a missionary described them as 'genuine Malthusians' (Munro & Bedford 1980, p. 3). Nearly a century later population densities were much higher, food shortages had occurred, poverty was common, and the colonial administration proceeded with resettlement. Two decades later the thousand Phoenix Islands settlers, and others from the Gilbert Islands, were relocated to the Solomon Islands, after recurrent drought in the Phoenix Islands. Moreover, the Phoenix Islands, while 'a healthy place to live', offered monotonous diets and 'little prospect for future development' (Knudson 1977, p. 214). Environmental threats were not the sole reason for resettlement.

The Gilbertese (i-Kiribati), who had previously settled in the Phoenix Islands, were relocated to the Solomon Islands, mainly around Gizo, in the mid-1950s (where, half a century later, they were some of the main victims of the 2007 tsunami). The Solomon Islands then appeared thinly populated. Vaitupu (Tuvalu) islanders were resettled on Kioa island in Fiji around the same time, although resettlement extended into the early 1980s (Koch 1978). This resettlement was not motivated by population pressure but rather seen as (literally) an investment against future overpopulation.

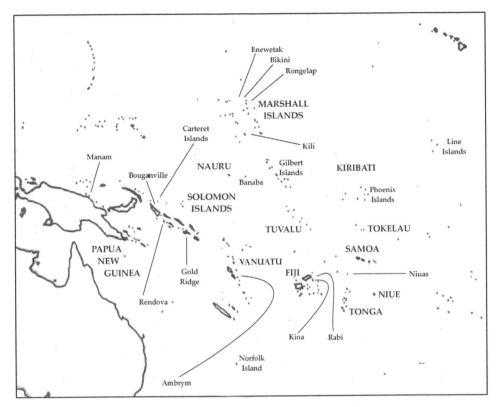

FIGURE 1. Resettlement Sites in the Pacific.

Such moves across what are now international borders are no longer regarded as politically feasible. Indeed, as early as the 1950s, a proposal to relocate other Tuvaluans in Tonga was rejected there since it was argued that the land would later be needed for Tongans (Connell 1983b). The process of resettlement, however, continued into post-colonial times in Kiribati with attempts from the early 1980s to resettle people within the country from the more densely populated Gilbert Islands chain to the northern Line Islands (and especially Kiritimati) 2500 km to the east. Inadequate infrastructure in the Line Islands and the great distance from the capital Tarawa made this unsustainable for a least-developed country, and migratory population targets were never met (ANU Enterprise Proprietary Ltd 2009).

A somewhat similar movement was instigated by New Zealand from the overcrowded atolls of Tokelau to New Zealand from the 1970s that resulted in Tokelauans becoming a majority overseas. Government-assisted migration to New Zealand began in 1963 and was explicitly justified as a response to 'future problems of overcrowding'. It was initially envisaged to eventually include the entire Tokelauan population. When the scheme was finally ended in 1976 some 528 people had been resettled under its auspices and many others had migrated freely (Wessen *et al.* 1992; Huntsman & Kalolo 2007). The Tokelau atoll population has subsequently stabilised and very slowly declined while the Tokelauan population has grown in New Zealand.

In each of these cases migration took islanders from relatively impoverished coral atolls, initially to atolls elsewhere, and subsequently to larger states and high islands. While the motivation was linked to overcrowding and lack of economic opportunities, as it continues to be (Shen & Gemenne 2011), such moves have some resonance and relevance for future contexts where sea level rise and its consequences are envisaged to pose substantial environmental hazards.

Environmental degradation? The Carteret Islands

The movement of Carteret Islanders to Bougainville, Papua New Guinea (PNG), also represents human-induced migration from 'overcrowded' atolls, despite its frequent attribution to climate-change-induced sea level rise. Indeed, Carteret Islanders have even been described as 'the first direct climate change refugees with islands inundated and damaged, gardens and water supplies destroyed by salt water intrusion and evacuation announced in 2005' (Roper 2009, p. 5). Numerous accounts of the contemporary Carteret Islands have suggested that environmental damage has resulted from global warming and sea level rise. However, records of sea levels in this region are non-existent. Damage has been caused by king tides, probably associated with particular El Niño conditions, and movement of nearby plate boundaries may also have had some effect. The island periodically experienced food shortages of varying severity, at least since the 1960s, alongside a steadily growing human and pig population. The increased conversion of coconuts into copra to generate income contributed to food shortages. Similar reports have traced such problems over several decades, and resettlement was considered at least as early as the 1960s, when islanders were reported to have a unanimous desire for resettlement, despite the idea being rejected on other Bougainville atolls (e.g. Mueller 1972; O'Collins 1990; Campbell 2010). An 'Atolls Resettlement Scheme' has been officially in operation since late 1984, with resettlement in Kuveria (Bougainville), on previously alienated government land. This appeared to begin well despite only a very small number of participating households (Kukang et al. 1987) but settlers had left the scheme by the late 1980s, even prior to the violent crisis which disrupted Bougainville for a decade. Most of those who returned went back because of frustration in acquiring title to land on Bougainville and because they 'were afraid they would lose their rights to the little land that was available on the islands and ultimately be left without land rights at all' (O'Collins 1990, p. 267). A further resettlement project was re-instigated in 2009 but various attempts to find appropriate land on Bougainville were thwarted by local landowners (Leckie 2009), and resettlement has largely failed. In 2011 it was nonetheless planned to resettle between 1350 and 1700 of the island's population—rather more than half of them—by 2020. By mid-2011 just two families had been resettled at Tinputz, on the east coast of Bougainville, with five more reported to be moving imminently, with perhaps three more by the end of the year; it was also intended to build 40 houses in 5 years (ABC 2 June 2011; Radio New Zealand International 12 September 2011). Difficulty in acquiring legal rights to land and opposition from local landowners prevented government attempts to secure adequate sites.

In some respects, despite the environmental problems in the Carterets, the outcome of this particular settlement scheme was a sequel to earlier schemes in both PNG and the Solomon Islands when people were resettled from more densely populated parts of the two countries, such as Simbu and Sepik in PNG, and

Malaita in the Solomon Islands, to less densely populated areas, notably New Britain island and Oro Province in PNG and parts of the Guadalcanal Plains in the Solomon Islands. In each case this was linked to the development of oil palm estates. None of these schemes were entirely successful, mainly because of disagreements over land tenure and land registration (especially as indigenous population growth occurred in destinations), differences of ethnicity and culture, and problems of access to off-farm employment. However, that in New Britain was successful enough for the initial settlers to be flooded with later arrivals from their home areas (Ploeg 1972; Hulme 1982; Heath 1979; Koczberski & Curry 2004). It is unlikely that similar schemes would now be attempted because of increased pressure on scarce agricultural land. That too is partly the situation that faces the Carteret Islanders.

Mining and the military

A series of rather less benevolent resettlements has occurred where islanders have been moved away from land more or less forcibly acquired for military and other purposes. This includes migration from islands such as Bikini (and its neighbours)—to make way for nuclear testing—and from different islands in Kwajalein atoll (Marshall Islands) for a US missile range. A series of short-distance migrations have followed the development of large mining projects, notably in Bougainville (PNG) and at Gold Ridge (Solomon Islands), while Banaba (Gilbert Islands/Kiribati) was depopulated to make way for phosphate mining.

As early as the late 1930s Japanese military forces had relocated the indigenous population of Kwajalein atoll (Marshall Islands) from the main islands to smaller islets in the atoll. At the start of its post-war weapons testing program, the USA once again removed Marshallese from the main island, and relocated them at high densities onto nearby Ebeye. In the 1940s islanders were moved from Bikini and Enewetak, and later Rongelap, in the northern Marshall Islands, to distant islands, notably Kili (then unpopulated) and the capital, Majuro (Kiste 1974). On Bikini, Rongelap and Enewetak land was the visible representation of centuries of human labour; for the Enewetakese 'in eating the products of their atoll ... people take on characteristics of that place' whereas eating food from 'foreign lands', even neighbouring Ujelang, was fraught with danger and uncertainty (Carucci 2004, p. 419). On Ujelang, the most remote Marshall Islands atoll, the Enewetakese experienced waves of famine, while they, the Bikinians and the Rongelapese (the last dispersed rather than settled in one place) simply yearned to return home (Carucci 2004; Kiste 1974, 1985; Barker 2004). Eventually Enewetakese did return but to land 'desiccated and unrecognizable' where they could not be at home 'in the very land that is their home' (Carucci 2004, p. 436). Such devastating destruction and neglect is relatively unusual.

Resettlement has also resulted from moving people away from large-scale mining activities, such as the Bougainville copper mine and the Banaba phosphate mine. In the former case migration was local, over a few kilometres, while in the latter case it was from the Gilbert Islands (now Kiribati) to Fiji. In Bougainville, resettled villagers resisted attempts to relocate them on resumed coastal plantations but preferred to remain on their own lands, despite poor access to resources. That choice posed problems in subsequent years as agricultural land was further alienated by the mine, and resettled villagers were dissatisfied with their new housing, compensation payments and continued environmental degradation (Connell 1991, 1992). Two decades later at Lihir island, despite landowners having larger houses, problems remained over access to adequate land, disputes over compensation and reluctance to move from sacred sites (Bainton 2010). At other large mine sites in upland PNG resettlement schemes have resulted in similar problems. The new Gold Ridge mine in the Solomon Islands necessitated the relocation of just over 2000 people scattered in 43 small villages (Australian Solomons Gold Limited and Gold Ridge Mining Limited 2009). The significant numbers involved, their diversity of livelihoods (some being subsistence farmers and others gold panners, with some having recently moved into the area), and the large number of villages suggest the potential for similar problems.

From the early twentieth century Banaba island was mined for phosphate. By the end of the 1930s it had become almost uninhabitable for the local population so that the British colonial government forced the resettlement of Banabans on Rabi (Fiji), an island already alienated for plantation use. Prior to the Banaban resettlement on Rabi, the island was privately owned and used as a copra plantation, until the British colonial government purchased the island (with phosphate royalties from Banaba). The Banabans came to Fiji in three major waves, with the first group of 703 arriving in 1945. Second and third waves came in the 1970s and 1980s, the latter following the end of phosphate mining in 1979. It was initially anticipated that they would return after two years and that their land rights on Banaba would not be compromised (Silverman 1971). Early on a group of about 60 people were repatriated to Banaba and the island population is now around 300. Rabi island, however, has a population of around 5000. In 2005 Fiji granted citizenship to the residents of Rabi and also Kioa. Although citizens of Fiji, Rabi islanders still hold Kiribati passports, remain the legal landowners of Banaba, and send one representative to the Kiribati parliament (a second one is elected in Banaba). They are also represented in the Fijian parliament. A major objective of the Banaban people is to secure the environmental restoration of Banaba, and enable a more significant return migration, but even if restoration occurred most would stay in Fiji.

Prior to independence lengthy discussions took place over the possibility of resettlement of Nauruans for the same reasons, as Nauru was mined away, and in the same way that Banabans had been resettled in Fiji. Plans were developed to resettle the Nauruans in Australia, or in one of Britain's Pacific possessions. Despite extended negotiations, the plans never eventuated. By 1964 the Nauruans had rejected, in turn, the idea of their dispersal as new citizens of either of three Western states (the UK, Australia or New Zealand), their resettlement as a community in a mainland enclave, or their resettlement as a separate community of 'New Nauru' on an island off the Queensland coast (Tabucanon & Opeskin 2011). Greater access to phosphate income after independence and the retention of coastal land ended further speculation about new settlement.

What was distinctive about resettlement in the two broad contexts of military exclusion (Bikini and neighbouring islands) and mining (Bougainville, Lihir, Gold Ridge and Banaba) was that migrants left reluctantly, with considerable trepidation, but were given no choice. Most sought to remain close to their land, experienced acute anxiety about its future and returned to it when they could. While the rationale for resettlement was different the outcomes have similarities, notably the impossibility of return, except thus far in the limited cases of Banaba and Enewetak.

Environmental hazard

In various places catastrophic environmental events have resulted in very necessary migration, usually anticipated to be short term, with migrants returning after the hazard event is over to resume normal lives, but perhaps in modified circumstances. The most dramatic of such migrations has been the abrupt depopulation of particular islands, where volcanic activity has occurred. However, there is both a continuum of impacts, from dramatic events to slow environmental degradation and change, and a parallel continuum between forced migration, essential after a dramatic event, and voluntary migration, where costs and benefits can be evaluated over time.

In Vanuatu volcanic eruptions on Lopevi resulted in its depopulation in 1960, with resettlement mainly on nearby Paama. Elsewhere in Vanuatu a small proportion of the population of Ambrym were resettled on the central island of Efate after a major eruption in 1950 (Tonkinson 1977, 1979). Volcanic eruptions on Niuafo'ou (Tonga) in 1946 prompted the resettlement of the population on larger islands to the south, though many were later to return (Rogers 1981). In 1981 Pagan (Northern Marianas) erupted and the population was resettled on Saipan. Except in Lopevi, at least some of these resettled populations invariably returned.

The largest recent migration resulting from volcanic activity has been that from Manam (Madang Province, PNG) where about 9000 Manam islanders were resettled on the nearby New Guinea mainland after devastating eruptions in 2004 in which five people died and the island became uninhabitable. Care centres were set up by NGOs on former plantation land rented by the provincial government. A Manam Resettlement Authority was created in 2006 but it neither eventuated nor functioned because of the misuse of funds (Memafu 2011). It was not until 2007 that the PNG government made a large area of the mainland available for permanent resettlement from Manam. Islanders cleared the land themselves, but were given little other government assistance, and were segregated from the mainland community, 'reducing their morale and leaving a feeling of worthlessness' (Mercer & Kelman 2010, p. 417). Islanders were also told that under no circumstances would the government provide any protection for those who returned to the island. Crime levels increased, as tension built between the islanders and mainland residents, and education levels decreased as islanders struggled to adapt on the mainland, where resources were scarce and access to employment very difficult. By 2009 some Manam islanders had returned to one home village because of the problems of adjustment on the mainland, believing that that they were better able to cope with the impact of volcanic activity than the effects of food shortages in the care centres.

Volcanic eruptions rarely devastate entire islands, and migration is the result of a range of additional factors. In some cases, as on Niuafo'ou, migration was in large part the outcome of dissatisfaction with the level of development and services, with volcanic activity simply the catalyst for migration. While almost the entire population was resettled on the larger southern island of 'Eua after 1946, a handful chose to remain. Other groups went back in 1958 but about two-thirds chose not to return (Rogers 1981). Similarly, for Ambrymese migrants resettled at Mele-Maat on the outer edge of the capital, Port Vila, after volcanic eruption, only some chose to return to Ambrym, with most preferring the better facilities of Efate island and, rather later, the social and economic benefits of being on the fringe of the capital city (Tonkinson 1979). Resettlement has invariably been a catalyst for some longer term migration, since displaced populations have never returned in their entirety. All hazard events lead to some more or less permanent migration.

Tsunamis have usually contributed to more localised resettlement, notably in Upolu (Samoa) where many survivors of the 2009 tsunami chose to move to higher ground inland. Pacific Islands have experienced two significant tsunamis in the past decade. The 2009 Samoan tsunami killed nearly 200 people, mainly in Samoa but also in American Samoa and in Niuatoputapu (Tonga). Parts of the western Solomon Islands were devastated in 2007 when a tsunami struck close to Gizo the provincial capital; more than 50 people were killed and several thousand displaced. A further tsunami early in 2010 in the western Solomon Islands was centred on Rendova island, destroying as many as 200 homes and leaving one-third of the island population homeless. In Samoa and Niuatoputapu only some islanders were resettled inland and that also failed to occur in the western Solomon Islands, through lack of resources (Madfis et al. 2010). Many sought to remain in coastal locations.

Numerous examples of localised migration occur where there has been flooding, cyclone damage or earthquake activity. In the western Solomon Islands a 2007 earthquake created problems of food security for people on Ranongga island, since landslides that accompanied the earthquake were more severe than from earlier earthquakes because more land had been cleared for agriculture as the population had grown. Previous earthquakes had resulted in the resettlement of villagers from more rugged areas to coastal locations and the 2007 earthquake resulted in similar interest, but intensified coastal land tenure problems partly stemming from greater population densities discouraged it (McDougall et al. 2008). Resettlement on the coast was no longer easily possible because of limited and contested access to land. Most hazard events, including cyclones, result in no more than temporary relocation, despite being catalysts for some permanent migration.

In atoll contexts localised migration is particularly difficult; even in earlier times when possibilities did exist it was never easy (Vayda 1958). Consequently, in various places, and especially atolls, villagers have either constructed seawalls or moved inland where this has been possible. Even in quite remote atolls such as the Carterets, concrete seawalls had been built by the 1980s. In Kiribati, where houses have been built further from the shoreline, but in an atoll context, the President has observed: 'We keep moving back from the shoreline. In a country like Kiribati, with very narrow islands, the room to move back is very limited' (quoted in Green 2010, p. 19; cf. Connell & Lea 1992). Local resettlement is almost impossible.

Environmental hazards, or even the threat of them, have resulted in both resettlement and migration. Invariably some fraction of this is permanent, despite the problems that often attend migration; indeed, migration would be more significant were it not for land issues, as in Ranongga and Manam. As in Niuatoputapu, migration is also a function of economic factors rather than purely the outcome of environmental factors, which may be catalyst rather than cause. Migration takes various forms that may optimise income-generating possibilities and enable economic resilience, but where possible it is usually either nearby and/or to urban centres. In certain distinctive circumstances, therefore, environmental hazards are one influence on migration which is otherwise multivariate and more like that of many other parts of the Pacific where economic factors are significant, and urbanisation is a common outcome (Connell 2006).

Consequences and outcomes

Resettlement in colonial times was relatively straightforward. Land was relatively easily alienated, former owners were displaced (if necessary) and settlers were established, sometimes, as at Rabi, with apparently secure titles to land. In any case, pressure on land was much less as populations were smaller and the drive for commodification less evident. What now seem the more successful examples of resettlement, such as Rabi, Kili and Mele-Maat, were established well before independence. While success is relative, some settlers succeeded, particularly relative to those who remained behind, in their integration with a more 'modern' economy. In Kiribati in 1979, the year of independence, the settlers in the Solomon Islands

> In earlier days ... were called the land-hungry people; they were the unfortunate ones who did not have sufficient land. Now our values have changed. Settling overseas, beyond the oceans of our islands, is something to be sought after. Why? Because our population is still growing ... those who have resettled will not face the problems that we will face in the future. So now many consider them, the resettled ones, as the fortunate ones and they consider us to be the unfortunate ones. (Schutz & Tenten 1979, p. 127)

That was prophetic in terms of the difficult future for Kiribati, and especially the Gilbert Islands chain, but their 'fortune' has never been clear cut, especially as a minority in the Solomon Islands without secure access to land. Nonetheless, few have ever returned from the Solomon Islands, just as few Tuvaluans have returned from Kioa and few Banabans from Rabi.

Elsewhere, a range of issues have restricted successful resettlement. Firstly, post-colonial governments have been without financial resources, management capacity, commitment and focus to develop effective settlement schemes or identify the need for them. In Manam, for example, major eruptions had occurred in 1937, 1957 (when the Australian administration evacuated everyone to the mainland for an extended period of time), 1992 and 1996, but the 2004 eruption left both the provincial and national governments without any plan for relocation. Very similar problems of planning and land acquisition plagued the resettlement of people from Rabaul, after the 1994 eruption, despite a long history of volcanic activity. Institutional weakness has prevented adequate planning and management, few resources were available for possible future eventualities, and disorganisation, lack of resources, corruption and fragmented responsibilities made success even more difficult.

Secondly, in most contexts local opposition and resentment have prevented effective resettlement. As early as 1946, when 1300 Niuafo'ouans arrived in the capital of Nuku'alofa, Tonga, to be resettled after the volcanic eruption on their home island, 'they were viewed with no small concern and hostility by local residents … [stemming] from the fear of increased pressure on limited resources, especially land and jobs, and from social prejudice' (Rogers 1981, p. 158). In new sites on 'Eua fights later erupted between the settlers and established residents; only after 18 years had the settlers 'muted *much* of the original criticism and animosity of their neighbours' (Rogers 1981, p. 160; my emphasis). In the 1960s, although there was no overt hostility between Gilbertese settlers and Solomon Islanders, 'there were feelings of resentment on the part of many Melanesians, who viewed the Gilbertese as having taken land and jobs that ought to belong to Melanesians' (Knudson 1977, p. 223). Opposition was most evident in Melanesia.

After independence, integration of the Gilbertese in the Solomon Islands and the Banabans in Fiji proved more difficult. Nauruans rejected resettlement since they had no wish to 'assimilate' elsewhere. Bougainvillean atoll residents had no desire to move if it meant losing their cultural identity by being 'swallowed up' on a much larger island (O'Collins 1990, p. 250). Movements across cultural boundaries were particularly difficult. Dissatisfaction of the migrants themselves—with new unfamiliar environments (climates, mountains, rivers, languages and diseases) and even the vastly different scale of their new 'homes', becoming a minority in a new land with few options for employment, poor access to service and simple homesickness, which became nostalgia—reduced the chances of success. Most were reluctant to move and preferred nearby relocation. The small numbers migrating limited the significance of the movement for the places from which people moved, notably the Gilbert Islands, and resettlement was costly.

Thirdly, underpinning local opposition to settlement was land ownership. For Manam, resettlement created conflicts with the local people over the use of resources such as land for gardening, water, materials to build houses and access to the sea, circumstances that challenged notions of food security, and violence ensued (Matbob 2007). So difficult has resettlement been that Manam islanders have attempted to relocate to Manam even while the volcano remains active, with ash falls damaging food gardens and polluting water supplies. Similar problems have occurred with resettlement of Carteret Islanders in Bougainville, and Ambrymese in Efate, despite some land used for resettlement having been previously alienated. Resettlement in rural areas, even where land is seemingly vacant, poses social and cultural problems.

Landowners were increasingly reluctant to cede land to others, however moral and worthy their claims, even when they shared kinship ties. In post-colonial times land boundaries have become frozen and land too valuable for most islanders to be willing to enable even displaced people to lease or purchase it. Land once alienated for such things as plantations has been resumed by the former owners rather than made available to governments for potential settlers. It is, after all, inherently implausible that settlers can take advantage of good, idle and accessible land with owners who are willing to lease or sell it (cf. Wilmsen *et al.* 2011). Even where resettlement has been highly localised and within the same cultural area, land transfers have proved complex and challenging. After the 2009 tsunami in Niuatoputapu (where nine people were killed and over 80 buildings destroyed) almost all residents chose to remain on Niuatoputapu. The Tongan government sought to resettle them (and the hospital and other public buildings) on higher ground more than 10 m above sea level, but many wished to remain on their own land close to the coast. Owners of upland areas were compensated for the loss of their land to resettlement by being offered land on the main island of Tongatapu.

Encouraged by a Niuas migrant in Australia, who estimated a very high value for the land, some residents initially held out for higher prices, but their demands were eventually quashed, by withdrawing their particular plots from the resettlement zone. Thus even in the more regulated land tenure context of Tonga, absentees and local residents have complicated efforts at resettlement, although most of those being settled were kin.

Fourthly, and consequently, recent resettlement has posed problems not only of access to land and marine resources (hence issues of food security) but also of access to other resources such as employment, education and health services, in an increasingly competitive arena where unemployment is rife. That was especially so in remote locations, where land was available, often remote from urban centres (even if not needed by local people at the time of settlement), such as the Phoenix Islands and later Rabi, Kioa and Kili. Even where settlers have had effectively unrestricted access to land and marine resources, as in Kili and Rabi, new husbandry techniques have sometimes proved difficult to acquire and cultures have reluctantly adapted.

In most cases resettlement was increasingly resented, even where it appeared necessary as in Manam and Banaba, both by the settlers and the local people. 'Success' occurred primarily where unpopulated islands existed, as in Fiji. In the particular circumstances of the Marshall Islands, alienation and forced migration were strongly resented, and return was sought in a strongly politicised context, a situation with acute parallels to broadly similar circumstances in the Chagos archipelago in the Indian Ocean (Vine 2009). Resettlement was extremely unwelcome, and while many were frustrated at the impossibility of moving back to contaminated (or evaporated) home islands, more Bikinians and Enewetakese have subsequently frequently moved, often first to Majuro and then to the USA. Those resettled from mining sites, whether on Bougainville or Banaba, were equally distraught about what had happened to their lands, which in some respects no longer actually existed.

The issues that surfaced for the Manam and Carterets resettlement represent the tensions that commonly exist between migrants and established populations, especially where resources (above all land, but also employment and various services) are valuable and in short supply, and government regulation and support is weak. In various other cases where resettlement has occurred from remote islands, whether because of natural hazard or population pressure, administering authorities and governments have sought some combination of making resettlement permanent (and discouraging return migration), and ensuring that the entire population leaves, because of the costs of providing social and physical infrastructure to remote places. This was early evident for such 'international' contexts as Tokelau (Connell 1983a; Wessen et al. 1992; Huntsman & Kalolo 2007) and has been more recently evident in the Northern Marianas, but where numbers have been extremely small. (The British government had similarly sought to deter return migration to the remote Atlantic colony of Tristan da Cunha after the 1961 volcanic eruption when the island population was relocated.) In Kiribati (then the Gilbert and Ellice Islands colony) the remote Phoenix Islands were settled in the 1930s but in the 1960s the entire population of about 1000 people was removed and resettled in the Solomon Islands after problems deriving from drought, social isolation, poor communications, inadequate market opportunities and costly administration, and the Phoenix Islands were never settled again. Occasionally the victims of disasters have considered the prospect of abandonment themselves. Cyclone Heta, which devastated Niue in January 2004, destroying almost a quarter of the houses on the island, prompted some thoughts about the permanent abandonment of the island (with the remaining 1500 population following so many of their kin to New Zealand), as had Cyclone Ofa two decades earlier (Barker 2000). The costs of resettlement, and of basic administration on remote islands with small populations, and the extremely limited prospects for commercial development, have challenged both resettlement and administration, and emphasise the problems of sustainable development (economically, socially and ecologically) on small remote islands. Although Niue, with its steadily declining population, has cautiously welcomed migrants from Tuvalu, in contemporary times the kinds of resettlement that were possible in colonial times (such as from the Gilbert Islands to the Solomon Islands and Fiji) are no longer possible, and global concerns now influence movements on this scale, but in a political climate that is not particularly propitious to refugees and migrants of any kind.

Conclusion

Wherever resettlement has occurred, social tensions have followed, from the earliest days of Pitcairn resettlement on Norfolk Island. Tensions have been greatest where migration has crossed cultural boundaries, which are often nearby in Melanesian, whether at Gold Ridge or Manam, and in post-colonial times, as land has become an even more valuable asset and government less autocratic and effective and more contested. Regional cultural and linguistic diversity cannot be wished away. Tensions particularly centre on access to resources, notably land, but also to health, education and employment. Unfamiliar environments, phenomena particularly severe for atoll populations moving to high islands, have challenged adaptation. Even moving to adjoining atolls, as at Ujelang and Nassau, still posed social and environmental problems (Carucci 2004; Vayda 1958). Land tenure had to adapt to new contexts (Kiste 1974; Tonkinson 1977, 1979). Even after generational shifts, security of tenure can be challenged. As populations have grown, land has become increasingly valuable and contested, even between kin, as on Niuatoputapu. Settlers remain 'outsiders' even over generations, while identities become ever more complex (Knudson 1977; Teiawa 2005). Consequently, settlers are increasingly more like refugees, dependent on some external support, and less likely to settle, either returning whence they have come (however difficult that may be) or moving onwards, often closer to urban centres: a deterritorialisation by displacement. Ironically a poverty of opportunity is seen in the Pacific as arising from isolation from markets, services and opportunities but resettlement creates a 'new poverty' where adequate land and services are absent.

Stability is difficult to achieve without resources, after the initial reconstruction capital is used up and focuses have shifted elsewhere, without access to services and without goodwill. Most Pacific Island governments do not have the will or capacity to develop, legislate for or administer effective resettlement policies in the face of local anxieties, tensions and hegemony over land. Because resettlement within countries has become particularly problematic, mainly because of land disputes, it will become increasingly difficult to find locations within the Pacific for people displaced by hazard, yet future hazards are certain. Resettlement as an adaptation to hazards, where relocation is impossible on land belonging to the displaced group, is unlikely to be successful within Pacific Island states, suggesting that, unless local circumstances change, future resettlement may need to be in metropolitan states on the fringes of the Pacific where land can be effectively acquired by governments. All but the smallest, most localised future resettlement moves, where land is generally owned by the resettling population, are likely to require an international focus.

Collective migration is not necessarily wholly distinct from other forms of migration, where economic influences are crucial. All migrants both seek and welcome superior economic opportunities (hence relocation in a marginal area is unattractive) while a proportion of those who have moved because of environmental hazards remain in destinations (but not usually places of resettlement) even after the hazard event is over, where those locations offer superior forms of income generation. That has largely resulted in urbanisation and a withdrawal from the margins.

Most resettlement has enhanced rather than diminished the retention of island identities in the face of difference, especially since human rights have rarely been pre-eminent in such moves, most evidently in the Marshall Islands where Bikinians developed a 'deeply felt identity as a victimised people' (Kiste 1985, p. 117). At the very least there is nostalgia for home, expressed in different ways, as in Rabi (Teiawa 2012). Where people have moved collectively and involuntarily, particularly in the face of pressures to leave, whether they are environmental or human, it is unsurprising that such migrants retain at least some desire to return home, however implausible this may be. In many cases because of environmental degradation, physical loss of land (as in mining) or high levels of contamination, return is often impractical or in degraded and deprived circumstances. Physical displacement is paralleled by livelihood displacement. Even so, some Banabans and many others have gone home, to reclaim land and identity, as best they can. This longing for home distinguishes so much involuntary and collective resettlement from individual and household migration.

Resettlement has been most successful when planned and directed by colonial powers, as in the movement of Tokelauans to New Zealand, when time, planning and resources have existed, though even then not without problems (Huntsman & Kalolo 2007). They have been particularly successful when a legal regime has established land rights, especially where these are protected by reserve powers in the constitution, as at Rabi and Kioa, a situation that is unusual and unlikely to recur. They have been least successful where responses to environmental hazards have necessitated speed, and where settlers have no means of adequate or acceptable social and economic development and have not been involved in decision making, and have effectively been disempowered, as at Manam. Many human-induced resettlements have been equally problematic, notably in the Marshall Islands. Resettlement was relatively successful in Tokelau and Tikopia, and for the early settlers from the Gilbert Islands (Kiribati) in the Solomon Islands, because it aligned with what islanders wanted to do, which usually meant moving nearby or to an urban centre. By contrast, resettlement in remote islands, such as the Phoenix Islands, Rabi and even Niue, sometimes in places from where 'local' people have earlier moved away, without access to employment and services and/or in different political jurisdictions, has marginalised settlers. Resettlement has too often equated with disempowerment and loss of autonomy.

Few such resettlements have been well documented, despite the probable need for future resettlement (e.g. Ferris *et al.* 2011) but, although many of the data reviewed here are fragmentary and even anecdotal, there is substantial consistency and hence some conclusions are evident. The clearest lessons from decades of resettlement in the Pacific are, firstly, that is often unwelcome, even in response to hazards, by settlers and by 'recipients'; secondly, that it cannot be accompanied effectively without substantial financial resources and management ability (especially for planning and service provision), and such resources have rarely been available, especially in the post-independence era; thirdly, that resettlement is not a 'one-off event', but has long-term implications and resonance for sovereignty, identity, employment, land tenure and access to services, all of which have bearings on ultimate success; and, fourthly, that social, physical and legal infrastructures are necessary for success. Physical resettlement must be inclusive, and accompanied by social and economic development, and that poses problems for small states with weak economies and diverse cultures.

REFERENCES

ANU ENTERPRISE PROPRIETARY LTD (2009) *Republic of Kiribati: integrated land and population development program on Kiritimati Island*, ADB Technical Assistance Consultant's Report, Canberra.

AUSTRALIAN SOLOMONS GOLD LIMITED AND GOLD RIDGE MINING LIMITED (2009) *Resettlement Action Plan 2009. Gold Ridge Gold Mine, Guadalcanal, Solomon Islands*, Brisbane.

BAINTON, N. (2010) *The Lihir destiny: cultural responses to mining in Melanesia*, ANU E Press, Canberra.

BARKER, H. (2004) *Bravo for the Marshallese: regaining control in a post-nuclear, post-colonial world*, Wadsworth, Belmont, CA.

BARKER, J. (2000) 'Hurricanes and socio-economic development on Niue Island', *Asia Pacific Viewpoint* 41(2), pp. 191–205.

BONNEMAISON, J. (1984) 'Social and cultural aspects of land tenure', in Larmour, P. (ed.) *Land tenure in Vanuatu*, University of the South Pacific, Suva, pp. 1–5.

CAMPBELL, J. (2010) 'Climate-induced community relocation in the Pacific: the meaning and importance of land', in McAdam, J. (ed.) *Climate change and displacement: multi-disciplinary perspectives*, Hart, Oxford, pp. 57–80.

CARUCCI, L. (2004) 'The transformation of person and place on Enewetak and Ujelang atoll', in Lockwood, V. (ed.) *Globalization and culture change in the Pacific Islands*, Prentice Hall, Upper Saddle River, NJ, pp. 414–38.

CONNELL, J. (1983a) *Migration, employment and development in the South Pacific. Country Report No. 18: Tokelau*, SPC, Noumea.

CONNELL, J. (1983b) *Migration, employment and development in the South Pacific. Country Report No. 19: Tuvalu*, SPC, Noumea.

CONNELL, J. (1991) 'Compensation and conflict: the Bougainville copper mine, Papua New Guinea', in Connell, J. & Howitt, R. (eds) *Mining and indigenous peoples in Australasia*, Sydney University Press, Sydney, pp. 55–76.

CONNELL, J. (1992) '"Logic is a capitalist cover-up": compensation and crisis in Bougainville, Papua New Guinea', in Henningham, S. & May, R.J. (eds) *Resources, development and politics in the Pacific Islands*, Crawford House Press, Bathurst, pp. 30–54.

CONNELL, J. (2006) 'Migration, dependency and inequality in the Pacific: old wine in bigger bottles?', in Firth, S. (ed.) *Globalisation and governance in the Pacific Islands*, ANU E-Press, Canberra, pp. 59–106.

CONNELL, J. & LEA, J. (1992) '"My country will not be there": global warming, development and the planning response in small island states', *Cities* 9(4), pp. 295–309.

EMERY, J. (1985) 'Norfolk Island: a place of extremist punishment, and *Fenua Maitai*', *Heritage Australia* 4(4), pp. 5–9.

FERRIS, E., CERNEA, M. & PETZ, D. (2011) *On the front line of climate change and displacement*, The Brookings Institution London School of Economics Project on Internal Displacement, Washington, DC.

GREEN, M. (2010) 'Tiny islands face change that is hard to believe in', *Sydney Morning Herald* 27 November, p. 19.

HEATH, I. (ed.) (1979) *Land research in Solomon Islands*, Ministry of Agriculture and Lands, Honiara.

HULME, D. (1982) 'Land settlement schemes in Papua New Guinea: an overview', *Land Reform* 1, pp. 21–42.

HUNTSMAN, J. & KALOLO, K. (2007) *The future of Tokelau: decolonising agendas 1975–2006*, Auckland University Press, Auckland.

KISTE, R. (1974) *The Bikinians: a study in forced migration*, Cummings, Menlo Park, CA.

KISTE, R. (1985) 'Identity and relocation: the Bikini case', *Pacific Viewpoint* 26(1), pp. 116–38.

KNUDSON, K. (1977) 'Sydney Island, Titiana, and Kamaleai: Southern Gilbertese in the Phoenix and Solomon Islands', in Lieber, M. (ed.) *Exiles and migrants in Oceania*, University Press of Hawaii, Honolulu, pp. 195–241.

KOCH, K.-F. (1978) *Logs in the current of the sea*, ANU Press, Canberra.

KOCZBERSKI, G. & CURRY, G. (2004) 'Divided communities and contested landscapes: mobility, development and shifting identities in migrant destination sites in Papua New Guinea', *Asia Pacific Viewpoint* 45(3), pp. 357–71.

KUKANG, T., SELWYN, J., SIAU, A., TADE, E. & WAIRIU, M. (1987) 'Atolls Resettlement Scheme, North Solomons Province', in O'Collins, M. (ed.) *Rapid rural appraisal: case studies of small farming systems*, Department of Anthropology and Sociology, University of Papua New Guinea, Port Moresby, pp. 65–89.

LECKIE, S. (2009) 'Climate-related disasters and displacement: homes for lost homes, lands for lost lands', in Guzman, J., Martine, G., McGranahan, G., Schensul, D. & Tacoli, C. (eds) *Population dynamics and climate change*, International Institute for Environment and Development, London, pp. 119–32.

MADFIS, J., MARTYRIS, D. & TRIPLEHORN, C. (2010) 'Emergency safe spaces in Haiti and the Solomon Islands', *Disasters* 34(3), pp. 845–64.

MATBOB, B. (2007) 'Politics: waiting and waiting three years after', *Islands Business* 33(7), p. 28.

MAUDE, H.E. (1968) *Of islands and men*, Oxford University Press, Melbourne.

MCDOUGALL, D., BARRY, I. & PIO, S. (2008) 'Disaster and recovery on Ranongga: six months after the earthquake in the western Solomons', mimeo, Perth.

MEMAFU, P. (2011) *Climate induced migration in Papua New Guinea*, National Disaster Management Office, Port Moresby.

MERCER, J. & KELMAN, I. (2010) 'Living alongside a volcano in Baliau, Papua New Guinea', *Disaster Prevention and Management* 19(4), pp. 412–22.

MUELLER, A. (1972) 'Notes on the Tulun or Carteret Islands', *Journal of the Papua New Guinea Society* 6(1), pp. 77–83.

MUNRO, D. & BEDFORD, R. (1980) 'Historical backgrounds', in *A report on the results of the census of Tuvalu*, Government of Tuvalu, Funafuti, pp. 1–13.

O'COLLINS, M. (1990) 'Carteret islanders at the Atolls Resettlement Scheme: a response to land loss and population growth', in Pernetta, J. & Hughes, P. (eds) *Implications of expected climate changes in the South Pacific region: an overview*, UNEP Regional Seas Reports and Studies No. 128, Nairobi, pp. 247–69.

PLOEG, A. (1972) 'Sociological aspects of Kapore Settlement', in *Hoskins development: the role of oil palm and timber*, New Guinea Research Bulletin No. 49, New Guinea Research Unit, Canberra, pp. 21–118.

ROGERS, G. (1981) 'The evacuation of Niuafo'ou: an outlier in the Kingdom of Tonga', *Journal of Pacific History* 16(3), pp. 149–63.

ROPER, T. (2009) 'Implications for the Pacific', *Tiempo* 70, pp. 3–7.

SCHUTZ, B. & TENTEN, R. (1979) 'Adjustment: problems of growth and change, 1892 to 1944', in Talu, A. *et al.* (eds) *Kiribati: aspects of history*, Institute of Pacific Studies, Suva, pp. 106–27.

SHEN, S. & GEMENNE, F. (2011) 'Contrasted views on environmental change and migration: the case of Tuvaluan migration to New Zealand', *International Migration* 49(S1), pp. 224–42.

SILVERMAN, M. (1971) *Disconcerting issue: meaning and struggle in a resettled Pacific community*, University of Chicago Press, Chicago.

TABUCANON, G.M. & OPESKIN, B. (2011) 'The resettlement of Nauruans in Australia: an early case of failed environmental migration', *Journal of Pacific History* 46(2), pp. 337–56.

TEIAWA, K. (2005) 'Our sea of phosphate: the diaspora of Ocean Island', in Harvey, G. & Thompson, C. (eds) *Indigenous diasporas and dislocations*, Ashgate, Aldershot, pp. 169–91.

TEIAWA, K. (2012) 'Choreographing difference: the (body) politics of Banaban dance', *The Contemporary Pacific* 24(1), pp. 65–94.

TONKINSON, R. (1977) 'The exploitation of ambiguity: a New Hebrides case', in Lieber, M. (ed.) *Exiles and migrants in Oceania*, University Press of Hawaii, Honolulu, pp. 269–95.

TONKINSON, R. (1979) 'The paradox of permanency in a resettled New Hebridean community', *Mass Emergencies* 4(2), pp. 105–16.

VAYDA, A. (1958) 'The Pukapukans on Nassau Island', *Journal of the Polynesian Society* 67(3), pp. 256–65.

VINE, D. (2009) *Island of shame: the secret history of the US military base on Diego Garcia*, Princeton University Press, Princeton and Oxford.

WESSEN, A., HOOPER, A., HUNTSMAN, J., PRIOR, I. & SALMOND, C. (1992) *Migration and health in a small society*, Clarendon Press, Oxford.

WILMSEN, B., WEBBER, M. & YUEFANG, D. (2011) 'Development for whom? Rural to urban resettlement at the Three Gorges Dam, China', *Asian Studies Review* 35(1), pp. 21–42.

Migration, Informal Urban Settlements and Non-market Land Transactions: a case study of Wewak, East Sepik Province, Papua New Guinea

GEORGINA NUMBASA & GINA KOCZBERSKI, *University of Papua New Guinea; Curtin University of Technology, Australia*

ABSTRACT   *This paper examines the various ways in which migrant settlers have gained and maintained access to land in the informal urban settlements of Wewak, the provincial capital of East Sepik Province, Papua New Guinea (PNG). Urban population growth in PNG and in Pacific Island states more generally is predicted to grow rapidly over the next two decades. Given the limited availability of formal housing for lower income people, it is likely that many will live in informal urban settlements on land owned by customary landowners. To date, there is very little information on how migrants living in informal settlements obtain and maintain access to land to erect dwellings and pursue livelihoods. Drawing on field research carried out in seven informal settlements in Wewak, the paper describes the historical, trading and/or marital ties between landowners and the original settler community. The discussion focuses on how access rights are maintained and have changed over time as the social and exchange relationships deteriorate between second-generation urban migrants and younger-generation landowners. The weakening of the social relationships between these two groups undermines the long-term use rights of migrants. By examining the changing tenure security of second-generation migrants the paper shows that whilst informal land markets perform an important role in housing provision for the urban poor they often fail to deliver long-term tenure security. The paper finishes with a brief consideration of the implications of the research findings for guiding policies on urban land reform and planning on customary land in PNG.*

Introduction

About 26 per cent of the estimated Pacific region population of 9.8 million resides in urban areas, with the largest proportions of urban populations established in Micronesia, followed by Polynesia and Melanesia (Jones 2011). Although urbanisation in the Pacific is a relatively recent phenomenon, all Pacific Island

29

states are rapidly urbanising, with urban growth rates outstripping national growth rates. In Papua New Guinea (PNG), it is likely, given present growth rates, that within the next 20 years Port Moresby's population will reach half a million, up from about 10 000 in the 1950s (Ward 1971) (see Figure 1). Growth projections for Greater Noumea indicate that the population will double in 20 years and for Port Vila it is estimated that population doubling time is 8 years (Storey 2003).

Managing urbanisation and urban growth will become one of the principal challenges for Pacific Island nations. Some of the most pressing urban planning issues facing Pacific towns and cities include: acute land shortages and conflicts; lack of affordable housing; the uncontrolled growth of informal settlements; income inequality and poverty; inadequate infrastructure and urban services; and environmental degradation (Storey 2006; Mawuli 2007; Connell 2011; Jones 2011). Among the most serious urban growth issues to be tackled will be to find effective ways to manage and facilitate access to land, primarily customary land, for housing and infrastructure to meet the needs of the rapidly growing number of people moving into urban areas and to cater for the natural growth in urban centres. Rapid urbanisation since the 1970s and the failure of many Pacific Island nations to successfully manage urban land have seen the spread of unplanned informal settlements in urban centres, especially in peri-urban areas. In several Pacific Island states, informal settlements grow at two or three times the rate of the urban whole (Storey 2006). In Port Moresby and Suva, the two largest cities in the Pacific region, informal settlements now house more than half of the urban population (Jones 2011), and in Vanuatu's urban centres of Port Vila and Luganville, informal settlements provide essentially the only affordable housing for many Ni-Vanuatu (Storey 2006).

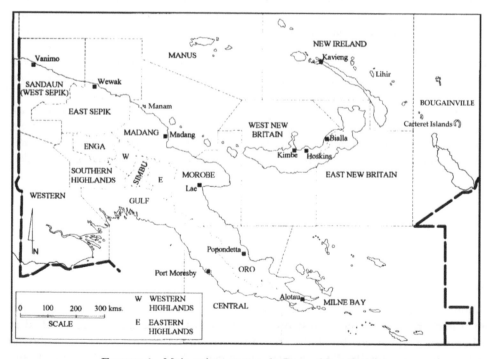

FIGURE 1. Main urban centres in Papua New Guinea.

Given that most of the future population growth in Pacific Island states will occur on customary land in urban and peri-urban informal settlements, it is vital for urban planners and policy makers to develop strategic plans that facilitate the mobilisation of customary land for future urban housing requirements while maintaining indigenous ownership and land tenure security for urban migrants. In much of the Pacific these issues have received little policy attention, although the release last year in PNG of the National Urbanisation Policy, to guide the management of urbanisation to 2030, clearly recognised the urgent need to address the permanency and growth of informal settlements in urban planning strategies and to work with customary landowners to mobilise their land for housing and urban development (Office of Urbanisation 2010).

This paper focuses on urban management issues in PNG. By 2030 it is estimated that approximately 35 per cent (3.5 million) of the population of PNG will be living in urban areas, nearly half of whom will be living in informal settlements (Storey 2010). One of the principal challenges for urban planners and managers, as identified by National Urbanisation Policy, will be finding ways to facilitate access to customary land for affordable housing and other urban infrastructure. About 97 per cent of land in PNG is held communally under customary tenure with access rights based on a mixture of descent, co-residence and participation in communal activities. State intervention is limited on customary land. The remaining land has been alienated from customary ownership and converted to state or private freehold title. PNG's urban centres have a mix of predominantly state land with scattered pockets of freehold and customary land. In Port Moresby, approximately 40 per cent of land is under customary tenure, and in the majority of cases migrants access customary land for housing through informal rental or land 'sale' agreements with customary landowners (Office of Urbanisation 2010). Much of the state and freehold land in urban areas which was not developed legally has been occupied by migrants and some unused state land has been informally reclaimed, or has been subject to compensation demands, by customary landowners (see Chand & Yala 2006). In Port Moresby, there are 20 planned settlements and 79 unplanned informal settlements[1] of which 42 are on state land and 37 on customary land (UN-Habitat 2010). Given the shortage of undeveloped state and freehold land, it is anticipated that the bulk of future urban growth will be on customary land (Office of Urbanisation 2010), and most migrants will live in informal peri-urban settlements on land leased or rented from customary landowners.

To date, the workings of the informal urban land sector and how migrants maintain long-term access rights to customary land in urban PNG have been neglected by urban planners. Yet formulating effective urban development plans that address the growth of informal settlements on customary land requires learning more about how informal settlements on customary land function and how migrants negotiate agreements with customary landowners to allow them long-term and secure access to customary land. This paper aims to address this gap in our knowledge by presenting results from fieldwork conducted in seven informal settlements in Wewak, East Sepik Province. The paper begins with a discussion of informal settlements in PNG before providing a brief background to Wewak and the results of the study which describe the social and economic lives of migrants and how they have gained and maintained tenure rights to land. The paper concludes by outlining the policy implications of the study for urban planning and urban land reform in PNG.

Informal settlements in Papua New Guinea

In all of PNG's major urban centres the migrant population typically comprises over half of the total urban population (National Statistical Office 2001). Many of those moving to urban centres settle in the expanding, unplanned urban and peri-urban settlements (Walsh 1987; Connell 1997; Koczberski *et al.* 2001; Office of Urbanisation 2010). In most urban centres, the growth and expansion of informal settlements have been on both state and customary land, with more recent growth located almost entirely on customary land in peri-urban areas. Jackson (1977) notes that the first informal settlements in Port Moresby appeared shortly after the Second World War. Initially, informal settlements in Port Moresby were established on customary land and most residents were Papuans who made arrangements with the customary landowners with whom they had some prior traditional trading relationships (Oram 1976; Hitchcock & Oram 1967). In the 1960s and 1970s, as informal settlements grew in size, personal exchange relationships with customary landowners weakened (Oram 1976; Koczberski *et al.* 2001; Goddard 2005). Also, the ethnic composition of informal settlements began to change from largely Papuan settlers, or what Goddard (2005) terms 'micro-ethnic enclaves' of people sharing a region of origin, to informal settlements housing migrants from other areas of PNG, especially the Highlands and coastal provinces, who lacked trading ties with the customary landowners.

The growth of informal settlements was linked to the easing of restrictions on the migration of the indigenous population into towns, as well as the rapid expansion of urban-based employment during the 1960s and 1970s and the increased government spending on welfare provision (Koczberski *et al.* 2001). During this period of rapid urban growth, the provision of housing for government and private-sector employees and the release of land for housing developments failed to keep pace with the influx of migrants. With no effective housing policies, the size and number of informal settlements grew substantially, with many forming on unoccupied state land (Jackson 1977; Walsh 1987; Connell 1997; Goddard 2005). In Port Moresby, for example, settlement populations during the period 1980–2000 grew at an annual average rate of 7.8 per cent, nearly double the growth rate of the National Capital District (Chand & Yala 2012). Despite very limited employment opportunities and housing, and an administrative reluctance to recognise the permanency of urban settlement, informal settlements have continued to grow in size and number and are now a prominent feature of PNG's urban landscape.

Since the 1990s, as urban problems have escalated, anti-urban sentiments and negative portrayals of informal settlements have intensified. Over the past decade informal settlement evictions and police raids have occurred in most major towns and cities (e.g. Port Moresby, Lae, Bulolo, Goroka, Rabaul, Kokopo and Wewak). Occasional public announcements by politicians that 'the government will no longer tolerate illegal settlement [*sic*] throughout the country' (Acting Prime Minister Dr Temu, quoted in *Post Courier* 20 May 2009, p. 5) remain prominent responses to the challenge of managing current and future demand for urban land for housing and urban services (Koczberski *et al.* 2001; Connell 2003, 2011; Goddard 2005). In the daily editorial of PNG's national newspaper, the editor noted in response to a wave of negative comments on informal settlements by politicians and threats of further settlement evictions in urban centres that:

> all governing authorities of Papua New Guinea, from the colonial administration of the 1960s and into the independent reign of our own people, have shirked the task of doing something constructive about settlements. The problem is still with us and will fester further ... Our national and provincial leaders seem incapable of devising sensible, humane plans to either improve the settlers' conditions or to find a way to establish decent housing estates for the hundreds of thousands in our towns and cities. (*Post Courier* online 21 May 2010)

Although there is little information on formal and informal activities in urban centres of PNG, it appears that the portrayal of informal settlements as hubs of the unemployed is not entirely correct. The reality is that with the severe shortages of formal housing in towns and cities in PNG, especially in Port Moresby, informal settlements have a wide socio-economic range of residents, with some working in the formal sector and the majority relying on the informal sector as their primary livelihood activity (Barber 2003; Umezaki & Ohtsuka 2003; Goddard 2005; Mawuli 2007; Chand & Yala 2012). In the informal settlement of Tari Market in Lae, where rooms are rented to meet the needs of the low-income workforce, one landlord estimated that almost half of the 700 residents were in formal waged employment (*Post Courier* 30 November 2009). With the current resources boom in PNG, and the pressure on housing in the nation's capital, it is likely that many in the formal waged sector have been forced into living in informal settlements. Indeed, in a graduation speech earlier this year, which made the front page of the national newspaper, the Vice Chancellor of the University of Natural Resources and Environment (UNRE) expressed his concerns about the poor working and housing conditions of young professional Papua New Guineans and said:

> I know of professionals, former students of UNRE and other universities that are still sleeping with *wantoks*, parents and relatives or are renting sub-standard accommodations in the Waigani swamps, Morata kunai or in Nuigo and Saksak compound and [in] many settlements that are springing up across the nation. (*Post Courier* 28 March 2011, pp. 1–2)

In acknowledging the critical housing shortages and social problems in urban centres, the National Urbanisation Policy identified the provision of 'an adequate supply and mix of land, that is customary, State and private' as the key policy area requiring urgent action (Office of Urbanisation 2010, p. 54). The report gave particular attention to mobilising undeveloped customary land. Such a policy priority is also in line with the 2007 National Land Development taskforce recommendation that land be made available for housing in urban areas through facilitating formal legal agreements with customary landowners, especially those in peri-urban areas (National Research Institute 2007).

Presently, the majority of urban migrants access customary land for residence through informal or semi-formal rental or land 'sale' agreements with customary landowners. These agreements are often based on kinship or friendship ties, or through long-standing traditional trading relationships with the customary landowning group (Oram 1976; Huber 1979; Norwood 1984; Goddard 2005; Chand & Yala 2012). The varied strategies that migrants have used to secure access to customary land for housing in Port Moresby have been documented by Chand and Yala (2012). They report that some migrant groups have signed formal statutory declarations validated by the Commissioner of Oaths to secure tenure rights to the land, and others retain 'rental' payment receipts to strengthen their occupancy rights. Those residing in Popondetta Settlement have established an institutional body (namely the Oro Development Community) to collect and record payments to the customary landowning group. In some situations, migrants have settled on customary land without approval of the customary landowning group or settled illegally on state land and are paying rental or other forms of payment to landowning groups claiming traditional ownership of the land (Chand & Yala 2012). Thus, migrants in informal settlements on customary or state land rely on a mix of informal and semi-formal agreements and social relationships to secure their access rights. The next section discusses the types of agreements migrants living in Wewak have pursued to maintain long-term access rights to land for housing.

Wewak and informal settlements

Wewak, on the north coast of PNG, is the provincial capital of East Sepik Province (see Figure 1) with a population of 20257[2] (National Statistical Office 2001). Wewak is the fifth largest urban centre in PNG and is connected to other areas of the province by the Sepik highway and its feeder roads, although over the last three decades the road networks throughout the province have deteriorated, leaving many areas inaccessible by road. The provision of basic goods and services to the bulk of the population is therefore very difficult, and this situation has fuelled rural–urban migration in the province. Many people in recent years, especially those from disadvantaged rural areas and the outer islands, have settled in Wewak to access services and livelihood opportunities.

After the Second World War through to the 1980s there was significant urban growth as Wewak developed as the administrative and commercial centre for the province, and service industries and other major infrastructure developments were established in the town. Owing to a lack of formal housing schemes for workers many accessed land from the customary landowners or the Catholic mission to build settlements. In the 1950s several informal settlements appeared on customary land and by 1974 almost 50 per cent of urban housing in Wewak was informal (Jackson 1977). Informal settlements continued to grow in the 1980s, and in 1987 a Catholic Church census recorded 29 ethnically based settlements in Wewak (Gewertz & Errington 1991). Thus, like other towns and cities in PNG, Wewak was becoming a town of migrants and by 2000 the migrant population of Wewak comprised 56 per cent of the total urban population (National Statistical Office 2001).

Accessing land

The process by which customary and Catholic mission land was accessed by migrants was investigated through interviews and structured household surveys in seven informal settlements.[3] The settlements included: Boram Beach Front, Kaindi Ward 5, Koil Island Camp–Kreer beach, Kuiya Settlement, Nuigo Settlement,

34

Mapau Settlement and Saksak Compound with an average household size of 6 (see Figure 2). The settlements occupied by people from nearby islands were located near the beach front (see Plate 1). The other settlements were located on swampy, waterlogged land, some of which were subject to regular flooding. For example, Kuiya Settlement is located on the banks of the Urarembe Creek and is subject to seasonal flooding, with its place name derived from the local language meaning 'coming waters'. The location of many of the settlements on swampy land, together with lack of proper drainage, garbage collection, water supply or sewerage, has created severe sanitation problems for residents (see Plate 2). Any suggested proposal that urban authorities will provide water supply, sanitation, garbage collection or drainage systems to the settlements is met with opposition from the landowners on which the settlements are located. Landowners argue that migrants are temporary residents and such services should be first provided to the villages where landowners reside.

Some residents were in wage employment, though most residents identified local marketing in the informal economy as their major source of income. This ranged from the sale of handicrafts and wood carvings to selling fish, store goods, cooked food and kerosene (see Plate 3). A few women were operating as informal money lenders. Most residents earned regular incomes, but many claimed that their income was barely sufficient to cover school fees, medical expenses, customary demands from village relatives and landowners, and the everyday expenses of urban living. Access to resources was also restricted in the settlements. Some settlers had made individual agreements with landowners to access land for gardening and some have gained access to land through marital ties with landowners. However, most

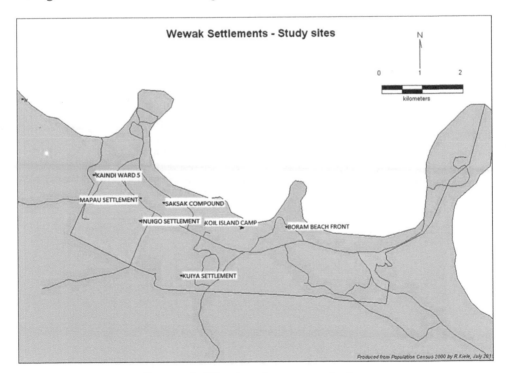

FIGURE 2. Wewak settlements: study sites.

35

PLATE 1. Koil Island Camp at Kreer Beach Front.

settlements did not have land available for food gardening as most land was taken up by housing or was too waterlogged and swampy. Overcrowding and unhygienic conditions together with limited livelihood options and crime made life difficult for many settlement residents. Thus, although better education and health services and income opportunities were available in Wewak compared with migrant source areas, life was not easy in the settlements.

The majority of the settlements were founded between 1950 and 1970; most of the older migrants have lived in their current settlements for over 30 years and

PLATE 2. Source of water supply at Mapau Settlement.

PLATE 3. Artefact sales at Saksak Compound.

were predominantly from the East Sepik Province (see Table 1). Some migrants first took up residence on what was then Catholic mission-owned land. The original landowners, many of whom were Catholic, supported the Church's activities and were therefore willing to give land to the Church, most of which was swampy and considered unproductive (e.g. Kuiya Settlement and Saksak Compound). Nuigo Settlement, the largest settlement in Wewak, was first established on mission-owned land but in the 1980s as the settlement grew the Catholic Church transferred the land to state title to enable the provision of municipal services such as water, sanitation and drainage. The mission land on which Kuiya Settlement and Saksak Compound were established has since been returned to the customary landowners. This has involved migrants creating new relationships with their new 'landlords'. Apart from Nuigo Settlement, all other settlements were located on customary land.

The process of establishing settlements on customary land varied, although all were initially established by a member of one ethnic group negotiating with the customary landowners for access to the land. Most were founded when male migrants came to Wewak seeking employment opportunities with the Catholic mission and the expanding government services, as well as to take advantage of education and health services that were not available in their home areas. In total, 42 per cent of residents surveyed in the settlements initially moved to Wewak to take up employment and a further 8.5 per cent accompanied a parent or relative who had employment in Wewak, or came independently seeking employment. Twenty-seven per cent of residents initially settled in Wewak to access health, education and other services. Some migrants were employed by the Catholic mission, and others settled on mission land because they had relatives at the settlement and/or were working in Wewak with no formal accommodation provided by their employers or were receiving medical treatment at the hospital. Once the initial male migrants were established in Wewak, they were joined by their

TABLE 1. Wewak settlement characteristics

Name of settlement	Dominant source area of migrants	Average No. of years living in settlement	Type of land tenure	Initial arrangement to access land	Informal arrangement with landowners to maintain access rights to land	Restrictions on residents by landowners
Boram Beach Front	Schouten Islands/ Murik Lakes	25	Customary	Traditional trading partners with landowners	Contribute in cash and kind to customary exchange demands	No fishing or hunting in mangroves or removal of material for house building or other purposes. Housing restrictions and control of settlement population
Kaindi Ward 5	Yangoru/west coast of Wewak	34.8	Customary (disputed)	Customary exchange	Contribute in cash and kind to customary exchange demands	Restrictions on fishing or hunting in mangroves and the removal of materials for housing. Housing restrictions and control of settlement population
Koil Island Camp	Schouten Islands	14.8	Customary	Traditional trading partners with landowners	Informal rents/ contribute in cash and kind to customary exchange demands	No small-scale business enterprises. Housing restrictions and control of settlement population
Kuiya Settlement	Torubu, Wosera, Drekikar, West Sepik	23.3	Customary/ some formerly mission land	Customary exchange/ mission land, compound for employees	Contribute in cash and kind to customary exchange demands	No large-scale business enterprises, no gardening, planting of cash crops or collecting firewood in adjoining land. Housing restrictions and control of settlement population
Nuigo Settlement	Middle Sepik River villages	24.4	State (formerly mission land)	Mission land, compound for employees	Formal rentals to national housing commission	No restrictions, although warned by neighbouring landowners not to take resources on their land
Saksak Compound	Sepik River villages	27.6	Customary (formerly mission land)	Mission land, compound for employees	Contribute in cash and kind to customary exchange demands	No restrictions
Mapau Settlement	Angoram	24.3	Customary	Mission land, compound for employees	Contribute in cash and kind to customary exchange demands	Informal tax paid on small businesses. Housing restrictions and control of settlement population

immediate and extended families. By developing along kin and clan lines, most settlements maintained the socio-cultural practices from their home villages and this formed the basis of the social structure in the newly formed settlements and helped maintain the cultural identities of each settlement. Over time and through intermarriage and expanding social and kinship networks, the ethnic composition of the settlements has become more varied.

In the post-war period, people from Wewak's offshore islands and distant coastal villages also began visiting Wewak more regularly for business, to visit relatives and to seek medical services and education. With the consent of the landowners, the migrants, many of whom were traditional trading partners, established small camps on the beach, such as Boram Beach Front and Koil Island Camp at Kreer beach. These camps were initially used by the visitors as places to secure their canoes when trading with their partners or visiting the mainland. In the case of Boram Beach Front and Koil Island Camp the initial negotiations were between elders from the migrant home villages and Wewak landowners. Later, agreements tended to be between individual migrants and customary landowners. Over time, as residency became more permanent, the camps developed into established settlements. These traditional trading ties were important for assisting later migrants to access land for more permanent settlement in Wewak.

Owing to the shortage of formal housing schemes, other migrants arriving in Wewak for employment, training or other purposes sought out local landowners and negotiated informal agreements and, in some cases, 'rentals' to establish housing. In many cases migrants who entered into customary arrangements with the landowners had some form of relationship with them, often based on friendship made through marital ties or some type of mutual interest. For example, at Kuiya Settlement, the migrants from Torubu originally gained access to the land through developing friendships with members of the landowning group. These friendships were strengthened through customary practices whereby ceremonial feasts were held for some migrants from Torubu and Aitipe who were adopted into the landowning clan and given local clan names.

The welcoming of migrants by some of the landowning groups in Wewak has also been reported by Huber (1979), who noted that landowners were eager to rent land to settlers, and many settlements were founded on agreements between the settlers and villagers to cooperate in commercial enterprises. These relationships were maintained through reciprocity, exchange partnerships and personal patronage. Thus in some instances Wewak landowners, and in particular village leaders, saw in their relationships with their traditional trading partners and allies opportunities to participate in the new monetised urban economy and enter into business collaborations through inviting them to settle on their land. In describing 'Kat', the village leader, who was central to inviting the Koil Islanders to settle on Kreer Village land, Huber notes:

> For Kat, modern conditions represented more than an opportunity for personal enrichment and the new migrants more than a larger labour force. Both provided an opportunity for him to perfect his performance as a *mandatua* [village leader/bigman] by simultaneously demonstrating his progressive ideas for production in a new environment and by expanding the numbers of people under his control and protection. (Huber 1979, p. 47)

Forty or fifty years since the settlements were established, migrants have continued to validate their tenure rights though personal exchange relationships with their customary landowning hosts and increasingly through intermarriage (see Table 1). When settlement residents living on customary land were asked what the land-owners expected from them to maintain their access rights, 74 per cent referred to regularly contributing in cash or kind to landowners' customary activities and expenses, especially funeral and brideprice expenses. Attending marriage ceremonies and other customary events in the landowning village (e.g. attending a funeral and providing food) were also considered important for building and strengthening their relationships with landowners. In some settlements, whilst informal rentals were initially paid to the landowners, only Koil Island Camp settlers have continued to pay rentals to the landowners. In other settlements, such as Kaindi Ward Camp and Saksak Compound, rental payments ceased when land disputes over the ownership of the settlement arose between subclans within the landowning group. Unlike some of the Port Moresby urban settlements studied by Chand and Yala (2012), none of the settlements in this study has developed institutions such as formal rent collection bodies, nor has any settler signed formal statutory de-clarations to secure permanent access to the land. Instead, migrants' in-kind and cash contributions towards the customary expenses of their landowning hosts have remained central to validating their settlement tenure rights. These payments forge ties between the two groups, and importantly publicly remind the landowners of migrants' respect for their customs and their ongoing commitment to honour the initial agreements made by the first migrants who established the settlements. Contributing to the customary expenses of landowners by settlement residents has been reported for other urban informal settlements in PNG (Hitchcock & Oram 1967; Levine & Levine 1979; Goddard 2005; Chand & Yala 2012) and also in settlements in rural locations (Curry & Koczberski 2009).

Ongoing access rights to land and housing in the settlements was also maintained through observing the restrictions placed on residents by the landowners and respecting their property rights (see Table 1). Landowner-imposed regulations varied amongst settlements and ranged from restricting the type of livelihood activities migrants could pursue, prohibiting the utilisation of mangrove or forest resources adjacent to settlements, and controlling house building and the numbers and demographic composition of the settler population. Nuigo Settlement being on state land had no restrictions on livelihood activities within its boundaries, however, the settlers were warned not to encroach onto neighbouring customary land for food gardening and collecting firewood. Only Koil Island Camp and Kuiya Settlement had restrictions on establishing small businesses, although livelihood opportunities were generally constrained in all the settlements owing to lack of available and/or suitable land for establishing business premises or cultivating food or cash crops. The most common landowner restrictions on settlements on customary land, apart from Saksak Compound, related to landowners determining who could reside in the settlement. In all cases, potential new residents had to obtain prior approval from the landowners and typically such approvals were given only to those who were closely related to a resident and who shared the ethnicity of the dominant ethnic group/s that were initially granted residence rights. Similarly, all new housing required prior approval before construction began, although given that most settlements were severely short of land there was very little area available for additional housing. Failure to observe these restrictions could lead to conflicts with landowners and possible evictions of settlers.

Despite the relatively stable relationships between landowners and settlers, an undercurrent of discontent was emerging which has undermined the tenure security of migrants in some settlements. Where different landowning groups were disputing the ownership of the land on which settlements were located, such as Saksak Compound and Kuiya Settlement, disgruntlement had arisen because some settlement residents had shown support and patronage to members of a landowning group who were not party to the original agreements between the settlers and village elders. Also, some settlers no longer fully abided by the restrictions agreed to in the original arrangements between the initial settlement residents and landowners. At Kuiya Settlement conflicts had erupted over the encroachment of settlers on to customary land to gather firewood, timber and coconuts. Moreover, some settlement residents had allowed relatives to settle with them without prior permission from the landowners, and this had undermined the tenure security of migrants. In an attempt to address the problem at Kuiya Settlement, residents formed a committee and set their own financial penalties for residents not observing the restrictions imposed by the customary landowners. By doing so, the committee publicly displayed to the landowners their respect for the original agreements made by their fathers and grandfathers. At some settlements people expressed their concern that younger residents did not show the same commitment as older residents to observing restrictions, and some settlers raised the problem of the younger members of the landowning group who likewise did not uphold or respect the original customary arrangements made with the migrants when they first settled the land.

Indeed, one of the more serious threats to tenure security stemmed largely from the changing relationships between younger members of the landowning group and the children born in the settlement who valued their exchange relationships and moral obligations with each other differently from their fathers/grandfathers. Amongst some of the younger and more recent settlers there was a perception that the financial demands and restrictions landowners placed on them were too onerous. Some younger residents resented complying with the endless requests to contribute cash to customary exchange of their landowner hosts, especially given their own household cash needs and their continuing customary obligations to relatives in their 'home' villages. Similarly, younger landowners no longer respected the mutual agreements or relationship ties established with the migrants by their fathers and grandfathers. Many argued that the agreements were of the past and belonged to an older generation for whom customary practices were more strongly observed. For these younger landowning members who no longer valued certain cultural practices and viewed migrants as the source of social problems and a hindrance to development, they preferred to see their land made available for formal leasing to developers. Landowners at Kuiya and Mapau Settlements and Saksak Compound had recently held discussions with developers about leasing their land.

One major ongoing source of discontent that threatened the long-term tenure security of migrants, and a topic raised frequently by informants, was the problem of drunkenness and drug abuse among settlement and village youth. Drug and alcohol abuse were seen to be associated with much of the petty crime, damage to property, harassment and street fights between youth from the settlements and landowning groups. Such behaviour has destabilised and weakened the social bonds between the two groups and undermined the occupancy rights of second-generation migrants. When serious conflicts arose between the two groups, leaders of the settlements would hold peace talks and make compensation payments to the landowning group to maintain their access rights to land. Thus, unlike land transactions in formal land markets, migrants' tenure rights are never guaranteed or permanent. Rather, migrants must validate their rights to the land of their hosts through ongoing maintenance of social and exchange relationships with them, while landowners themselves also regularly evaluated and reassessed the moral basis of migrants' occupancy rights. The non-binding nature of access rights placed second-generation migrants in a particularly vulnerable position as their personal exchange relationships with the landowners were changing and the importance of long-established marital and trading ties as a basis of migrant occupancy rights were given less recognition by younger members of the landowning group.

How second-generation migrants managed their relationships with customary landowners was therefore critical to determining their tenure status. Tenure security was especially important for the children born and raised in the settlements. Like second-generation migrants living on the land settlements in West New Britain (Koczberski *et al.* 2012), most second-generation residents were born in the settlements and almost 30 per cent of them were not fluent in their parents' home languages. Most settlement residents preferred the benefits of town living compared with village life, and many had no desire to return to live in their relatively isolated villages where health and education services and income opportunities were restricted, and where land shortages, land disputes and sorcery were ever-present threats (see Table 2). Although only 20 per cent of residents claimed that their children had lost access to land and other resources in their home villages, and despite the large majority of them having maintained strong ties with village relatives through regularly hosting visitors, remitting money home for customary activities and occasionally making brief visits to their village, most settlement

TABLE 2. Residents' perceptions of settlement and village life

Settlement and village life	Perceptions	%
What aspects of your life are different here (settlement) compared with your home village?	1. Better employment opportunities	9.2
	2. Better access to health and education services	44.4
	3. Better access to services and income opportunities	35.2
	4. Better access to land and other resources in home village	11.1
Is life better or safer living in the settlement than in your home village?	1. Settlement safer. In village, sorcery is a threat	31.7
	2. Settlement better. In village, no services	14.2
	3. Settlement life is better	6.3
	4. Village. Easier life and settlement not safe	28.5
	5. Both village and settlement not safe	11.1
	6. Both village and settlement safe	4.7
	7. No choice but to live in settlement	3.1
Where would you move if you had to leave the settlement?	1. Return to village	38.7
	2. Move to another settlement/area in town	35.4
	3. Move to another village/area/province in PNG	14.5
	4. Uncertain where to move	11.2

residents did not see returning to their village as an option if they were evicted from the settlements (see Table 2). Rather, living in another Wewak settlement or elsewhere in the province or PNG were considered to be better options. Gewertz and Errington (1991) also note that most of the Chambri migrants living at Chambri Camp Settlement in Wewak had no intention of returning to their home villages.

To avoid becoming dispossessed urban migrants, it is in the interests of second-generation settlers to maintain good and stable social relationships with their landowning hosts. As indicated above, these relationships were becoming strained with the younger generation, and it is possible that as more landowners recognised the economic potential of their land, especially portions of land located in prime areas for development, they would abandon long-standing land use agreements with migrants in favour of better returns. If second-generation migrants were to be evicted they would have to return to their home village or seek residence elsewhere in Wewak. Both options would present difficulties for migrants. Despite second-generation migrants having maintained ties with village relatives, it is likely that many of them have lost access to land and resources, and most would be incapable of defending their claim in village land disputes and mediation because they lack sufficient knowledge of their oral histories or genealogy, and may find relatives unwilling to support them. Studies elsewhere in PNG have also indicated that the maintenance of exchange relationships with home may not always be sufficient for an unproblematic re-entry to village society and the successful re-activation of resource rights (e.g. Morauta & Ryan 1982; Carrier & Carrier 1989; Neumann 1997). Zimmer-Tamakoshi (1997) reports how a Gende migrant returned home after 13 years' absence to find that he did not receive support in a land dispute, despite hosting his uncles for months at a time when they visited. The return migrant mistakenly believed that his earlier generosity in hosting visiting relatives would partly repay his debts to his uncles: they, on the other hand, only remembered his inadequate repayments on his brideprice debts for his first wife. For second-generation migrants who have lost their tenure rights to village resources, moving to another settlement in Wewak if forced to leave their current location would not be easy given the limits imposed by landowners on the population size and membership of informal settlements on their land. These migrants face a precarious future.

Discussion and conclusions

As identified in the 2010 National Urbanisation Policy, one of the major urban challenges in PNG will be to manage current and future demand for customary land for housing and services for the growing number of poor urban dwellers. As shown in this study, many of the Wewak settlements were initially founded in the 1960s and 1970s when employment opportunities were expanding and housing and land options were limited. Since then the urban population has grown significantly, yet affordable formal housing for lower income households and employment opportunities has remained limited. The result is that most urban migrants have been forced into poorly serviced informal settlements. At the same time many of the remote areas of the province have continued to perform poorly on key social indicators, and living standards and basic social service provision have hardly improved over the last two decades. While these rural–urban inequalities persist and land and formal housing remain in short supply, unplanned urban informal settlements on customary land will continue to grow in number and size for the foreseeable future.

In Wewak the informal land market that has developed is characterised by a range of informal tenure arrangements between migrant settlers and customary land-owners, with many initial residents involved in negotiating the agreement having some prior traditional trading relationship, marriage or friendship ties with the landowning group. Such ties aided their settlement. However, migrants' ongoing occupancy rights are mediated by place-based frameworks of land tenure in which personal exchange relationships and a moral economy of respect and obligations towards landowners play critical roles. Migrants' in-kind and cash contributions to landowners for customary expenses to validate their claims to land have also been reported from other urban and rural settlements in PNG (see Hitchcock & Oram 1967; Levine & Levine 1979; Goddard 2005; Curry & Koczberski 2009), with some urban settlements also including more formal systems such as rentals, statutory declarations or legally backed contractual arrangements to secure oc-cupancy rights (Mawuli 2007; Chand & Yala 2012). These diverse arrangements indicate that customary landowners, in most cases, are successfully modifying customary land practices to respond to the demand for land and housing in urban areas outside of government structures and formal land markets.

Whilst informal land markets perform an important role in housing provision for the urban poor, they often fail to deliver long-term tenure security for migrants. Deteriorating social relationships between landowning groups and second-generation migrants and the changing resident composition occurring in some of the Wewak settlements are undermining the long-term residency rights of migrants that have existed for decades. What worked well in the past when fewer migrants occupied the settlements and when most had a direct personal relation-ship with the clan members involved in the initial negotiations and agreements is under pressure in the contemporary urban context where social relationships between the two groups are waning, economic livelihood options are constrained and crime is a source of ongoing tension and mistrust. The changing relationship between migrants and landowners is also occurring at a time when segmentation amongst landowners and competing land claims are occurring, and this has the potential to further undermine residency rights for some migrants. Similar changing relationships between migrants and landowners have been reported in other urban centres of PNG and leave migrants vulnerable to eviction and harassment (see Koczberski *et al.* 2001; Connell & Lea 2002; Goddard 2005; Office of Urbanisation 2010). Land and social conflicts, settlement evictions and a growing intolerance of migrants in urban centres all point to the changing relationships between landowners and migrants and the increasing pressure on customary land in urban centres.

These trends suggest that in the changing urban environment a more formal land transfer system is required that identifies and 'registers' the owners of land, and puts in place a legal contractual lease agreement tailored to the contemporary urban situation that ensures landowning groups remain the 'owners' of the land and settlers have the necessary long-term security of tenure for housing and to pursue economic livelihoods. Ideally, the aim should be to complement and build on existing informal institutions and land transaction agreements that retain cus-tomary ownership, and develop procedures and practices in consultation with landowning groups, local government and migrants which involve some indigenous notions of land tenure and provide tenure security to both groups. Many of the failed attempts at land reform in PNG have tended to de-contextualise and de-historicise existing land transfer practices and land tenure principles, and instead have looked elsewhere to formulate policy. The informal land transactions operating in Wewak and in other urban and rural areas in PNG provide possible pathways for urban planners and managers tasked with addressing ways to mobilise customary land for housing and infrastructure (see Chand & Yala 2012; Koczberski *et al.* 2012).

However, urban land reform to mobilise customary land for housing must occur alongside wider efforts to improve urban governance and the institutional framework for urban management. Weak urban and national governance has resulted in very little progress in addressing the critical housing shortages for low income earners and resolving long-running land tenure disputes in urban centres. Customary landowners are key stakeholders in the urban context and their cooperation and collaboration with the state and local government in strategic planning is critical to improving housing availability, infrastructure development and the delivery of services. It has been argued that building relationships and partnerships between citizens and national and local government authorities that foster consultation and the participation of landowners and urban migrants in urban planning are central to meeting local needs and priorities and to better manage urban development, informal settlements and poverty (Storey 2006; Connell 2011). Unless more efforts are made to incorporate landowners as partners in urban management and planning, it will mean more of the same, that is, unplanned informal settlements, social conflicts and missed opportunities for economic development.

Acknowledgements

The Wewak research was part a larger ARC Discovery Project grant and received additional funding support through a Curtin University internal research grant. The assistance during fieldwork of students from the University of Papua New Guinea, George Numbasa and the Wewak community is gratefully acknowledged. We are grateful for feedback from the referees and George Curry.

NOTES

[1] The use of the term informal settlements in this paper follows the UN-Habitat Programme definition as encompassing those unplanned residential areas developed on land where the occupants have no formal title or occupy it illegally. In PNG, informal settlements are characterised by housing that does not comply with planning and building regulations, together with poor infrastructure and a lack of urban services.
[2] The census does not count the population living in 'villages' within the Wewak Urban LLG. Villages within the town boundary are counted under the Wewak Rural LLG.

[3] Fieldwork was conducted over a 2-week period in 2008 with the assistance of students from the University of Papua New Guinea. In total 60 structured household questionnaires and informal interviews were conducted. The survey questionnaire and interviews sought household information on population and demography, income, reasons for migrating, settlement residency history, tenure security, access to resources in the settlement and in their home villages, and maintenance of access rights to settlement land.

REFERENCES

BARBER, K. (2003) 'The Bugiau community of Eight-mile: an urban settlement in Port Moresby, Papua New Guinea', *Oceania* 73(4), pp. 287–97.

CARRIER, J.G. & CARRIER, A.H. (1989) *Wage, trade, and exchange in Melanesia: a Manus society in the modern state*, University of California Press, Berkeley.

CHAND, S. & YALA, C. (2006) 'Improving access to land within the settlements of Port Moresby', IDEC Working Paper 07/04, Crawford School of Economics and Government, Australian National University, Canberra.

CHAND, S. & YALA, C. (2012) 'Institutions for improving access to land for settler-housing: evidence from Papua New Guinea', *Land Policy* 29(1), pp. 143–53.

CONNELL, J. (1997) *Papua New Guinea: the struggle for development*, Routledge, London.

CONNELL, J. (2003) 'Regulations of space in the contemporary postcolonial Pacific city: Port Moresby and Suva', *Asia Pacific Viewpoint* 44(3), pp. 243–57.

CONNELL, J. (2011) 'Elephants in the Pacific? Pacific urbanisation and its discontents', *Asia Pacific Viewpoint* 52(2), pp. 121–35.

CONNELL, J. & LEA, J. (2002) *Urbanisation in the Island Pacific*, Routledge, London.

CURRY, G.N. & KOCZBERSKI, G. (2009) 'Finding common ground: relational concepts of land tenure and economy in the oil palm frontier of Papua New Guinea', *Geographical Journal* 175(2), pp. 98–111.

GEWERTZ, D.B. & ERRINGTON, F.K. (1991) *Twisted histories, altered context: representing the Chambri in a world system*, Cambridge University Press, Cambridge.

GODDARD, M. (2005) *The unseen city: anthropological perspectives on Port Moresby, Papua New Guinea*, Pandanus Books, Research School of Pacific and Asian Studies, Australian National University, Canberra.

HITCHCOCK, N.E. & ORAM, N. (1967) *Rabia Camp: a Port Moresby migrant settlement*, New Guinea Research Bulletin No. 14, New Guinea Research Unit, Canberra.

HUBER, M. (1979) 'Big men and partners: the development of urban migrant communities at Kreer beach', *Yagl Ambu* 6, pp. 39–49.

JACKSON, R. (1977) 'The growth, nature and future prospects of informal settlements in Papua New Guinea', *Pacific Viewpoint* 18(1), pp. 22–42.

JONES, P. (2011) 'Urbanisation in the Pacific Island context', *Development Bulletin* 74, pp. 93–7.

KOCZBERSKI, G., CURRY, G. & ANJEN, J. (2012) 'Changing land tenure and informal land markets in the oil palm frontier regions of Papua New Guinea: the challenge for land reform', *Australian Geographer* this issue.

KOCZBERSKI, G., CURRY, G.N. & CONNELL, J. (2001) 'Full circle or spiralling out of control? State violence and the control of urbanisation in Papua New Guinea', *Urban Studies* 38(11), pp. 2017–36.

LEVINE, H.B. & LEVINE, M.W. (1979) *Urbanisation in Papua New Guinea: a study of ambivalent townsmen*, Cambridge University Press, Cambridge.

MAWULI, A. (2007) 'Urban squatter settlers' livelihood in the National Capital District: case studies', in Mawuli, A. & Guy, R. (eds) *Informal safety nets: support systems of social and economic hardships in Papua New Guinea*, National Research Institute Publication No. 46, National Research Institute, Boroko.

MORAUTA, L. & RYAN, D. (1982) 'From temporary to permanent townsmen: migrants from the Malalaua District, Papua New Guinea', *Oceania* 53(1), pp. 39–55.

NATIONAL RESEARCH INSTITUTE (2007) *The National Land Development National Taskforce report: land administration, land dispute settlement and customary land development,*

Monograph 39, report prepared by the NLDT Committees on Land Administration, Land Dispute Settlement, and Customary Land Development, Boroko.

NATIONAL STATISTICAL OFFICE (2001) *National population and housing census 2000*, National Statistical Office, Port Moresby.

NEUMANN, K. (1997) 'Nostalgia for Rabaul', *Oceania* 67(3), pp. 177–93.

NORWOOD, H. (1984) *Port Moresby: urban villages and squatter settlements. An analysis of the urban villages and squatter areas of the city of Port Moresby in Papua New Guinea*, University of Papua New Guinea, Port Moresby.

OFFICE OF URBANISATION (2010) *National Urbanisation Policy for Papua New Guinea, 2010–2030*, Independent State of Papua New Guinea, Port Moresby.

ORAM, N.D. (1976) *Colonial town to Melanesian city: Port Moresby 1884–1974*, Australian National University, Canberra.

POST COURIER (2009) 'Illegal squatting not good', 20 May, p. 5.

POST COURIER (2009) '"Market" an unlikely haven', 30 November, p. 5.

POST COURIER (2010) 'Settlements: Govt must do something', 20 May. http://www. postcourier.com.pg (accessed 24 May 2010).

POST COURIER (2011) 'Uni graduates in squatters', 28 March, pp. 1–2.

STOREY, D. (2003) 'The peri-urban Pacific: from exclusive to inclusive cities', *Asia Pacific Viewpoint* 44(3), pp. 257–79.

STOREY, D. (2006) *Urbanisation in the Pacific*, State, Society and Governance in Melanesia Project, Research Paper, Australian National University, Canberra.

STOREY, D. (2010) 'Urban poverty in Papua New Guinea', Discussion Paper No. 109, National Research Institute, Port Moresby.

UMEZAKI, M. & OHTSUKA, R. (2003) 'Adaptive strategies of Highlands-origin migrant settlers in Port Moresby, Papua New Guinea', *Human Ecology* 31(1), pp. 3–25.

UN-HABITAT (2010) *Papua New Guinea: Port Moresby profile*, United Nations Human Settlement Programme, Nairobi, Kenya.

WALSH, B. (1987) 'The growth and development of squatter settlements in Lae, Papua New Guinea', in Mason, L. & Hereniko, P. (eds) *In search of a home*, Institute of Pacific Studies, University of the South Pacific, Suva, pp. 73–197.

WARD, R.G. (1971) 'Internal migration and urbanisation in Papua New Guinea', *New Guinea Research Bulletin* 42, pp. 81–107.

ZIMMER-TAMAKOSHI, L. (1997) 'Everyone (or no one) a winner: gender compensation ethics and practices', in Toft, S. (ed.) *Compensation for resource development in Papua New Guinea*, Law Reform Commission of Papua New Guinea and Resource Development in Asia and the Pacific, Research School of Pacific and Asian Studies, Australian National University, Canberra, pp. 66–83.

Land, Identity and Conflict on Guadalcanal, Solomon Islands

MATTHEW G. ALLEN, *Crawford School of Economics and Government, College of Asia and the Pacific, Australian National University, Australia*

ABSTRACT *This paper examines the recent history of settlement and conflict on Guadalcanal in Solomon Islands and the corollary emergence of competing 'settler' and 'landowner' identity narratives. Settlers from the island of Malaita were initially able to obtain rights to use customary land on north Guadalcanal but subsequently fell victim to a Guale project of exclusion. This project was informed by the breakdown of social relationships—both between landowners and settlers and within landowning groups themselves—and sharpening socio-economic differentiation amongst Guales along a north–south axis which has seen those from the relatively deprived south coast emerge as important actors. Localised grievances around land and settlement have been mobilised to the larger project of autonomy for Guadalcanal driven in part by a desire to capture a greater share of the benefits that flow from the island's resource industries. Against this backdrop of rapid socio-economic change, two competing identity narratives have emerged—a Malaitan settler narrative and a Guadalcanal landowner narrative—each of which seeks to establish a morally legitimate claim to property rights and economic opportunities on Guadalcanal. I provide windows onto each of these identity narratives and explore their articulation with historical and contemporary struggles over land and resource development.*

Sites of contestation over resource access and control in postcolonial Melanesia have been characterised by dynamic politics of identity as groups and individuals attempt to situate themselves differentially in relation to property and resource rights. This positioning is frequently pursued through projects of exclusion. The discursive simplification of customary tenure rendered as 'indigenous essentialism' (Hviding 1993), the deliberate truncation of the social networks which underpin belonging (Bainton 2009), and the construction of competing landowner and settler identity narratives (Koczberski & Curry 2004) are all indicative of the ubiquitous emergence of an 'ideology of landownership' that seeks to limit the claims of others (Filer 1997).

Similar processes have characterised resource struggles in neighbouring Indonesia where projects of exclusion have harnessed 'indigeneity' as their most potent trope. Indigeneity is deployed both as a symbol of resistance to the forces of globalisation and state appropriation, and as a means of securing or maximising the economic and livelihood benefits derived from land and resources (Li 2000; Elmhirst 2001). The rise of indigeneity has occurred alongside a revival, amongst theorists and activists alike, of place as an important concept in material and conceptual contests around globalisation and capitalist resource exploitation (Escobar 2001).

Most sites of resource conflict in Melanesia have featured 'outsiders', variously present as migrant workers or settlers, as important actors (Banks 2008). These groups are especially vulnerable to the projects of exclusion that attend the struggle for development in the context of rapid socio-economic change (Koczberski & Curry 2004). In Solomon Islands, the focus of this paper, the 'ethnic tension' of 1998–2000 saw the violent eviction of about 35 000 people from their homes on the island of Guadalcanal, many of whom were first-, second- or third-generation settlers from the neighbouring and densely populated island of Malaita.

It is not my purpose here to assess the salience of internal migration and settlement relative to other causes of the ethnic tension. Instead, I take as my subjects the process of settlement itself, in particular the means by which outsiders have obtained access to customary land on north Guadalcanal, and the emergence over time of a Guale (meaning 'indigenous' person(s) of Guadalcanal) project of exclusion aimed at limiting the land rights of settlers and challenging the legitimacy of the state in matters of resource access and control.

I highlight two processes that have informed this project of exclusion. First is the gradual breakdown of the social relationships—both between landowning groups and settlers and within landowning groups themselves—that have underpinned settlement on customary land on Guadalcanal. Second is the increasing socio-economic differentiation amongst Guales along a critical north–south axis which has seen those from the relatively deprived south coast, or Weather Coast, emerge as vocal and, during the ethnic tension, militant opponents of the presence of outsiders on 'their' island. Both processes have played out against a backdrop of rapid population growth and resource development on the northern side of the island since the Second World War, particularly in the areas immediately east and west of the capital Honiara.

With stark parallels to the situation in the oil palm regions of West New Britain Province in Papua New Guinea (Koczberski & Curry 2004), a Guale 'landowner' narrative invokes indigeneity as the paramount fount of legitimacy in the spheres of land and resource development, while a Malaitan 'settler' narrative emphasises nation building and national development, thereby linking itself to the legitimacy of the state. I employ interview material and published sources to provide windows onto each of these identity narratives and explore their articulation with historical and contemporary struggles over land and resource development.[1]

As has occurred in other Melanesian sites of enclave development, the politics of exclusion in this case have been mobilised to a larger project of 'statehood' for Guadalcanal, itself driven, *inter alia*, by a desire to capture a greater share of the benefits that flow from resource industries on Guadalcanal. It is in this arena that the Guale identity narrative has been cast most strongly in terms of indigeneity as it seeks to restrict the rights of outsiders, including the state.

I foreground the discussion by demonstrating how the 'Malaita' and 'Guadalcanal' identity narratives have been shaped by the economic history of the archipelago since the commencement of sustained contact with the global political economy from around the mid-1800s. Put very simply, this history has seen the 'under-development' of Malaita and 'over-development' of Guadalcanal (with the very important exception of the remote and impoverished Weather Coast) (Moore 2007, p. 231). The geographical pattern of uneven development has been the primary driver of internal migration.

In the final part of the paper I suggest that resource struggles in Solomon Islands have not been concerned solely with accessing, capturing or maximising the livelihood and economic benefits associated with land and resource development. They have been as much about controlling the nature of development itself. Not all Solomon Islanders have wanted commercial logging or mining or conservation projects on their land (Hviding 1993, 2003; Scales 2003). Many desire forms of development which are more attuned to the complexities of local land tenure systems and 'traditional' socio-economies (Gegeo 1998; Moore 2007). In this conceptualisation, indigenous projects of exclusion are better seen in terms of resistance rather than political economy or cultural difference.

Solomon Islands

Solomon Islands is an archipelago of over 1000 islands with a population of about 500 000 people distributed over six main islands and speaking around 80 languages. While the island of Malaita accounts for one-third of the total population, the national capital, Honiara, is on Guadalcanal where it was established immediately after the Second World War. About 85 per cent of Solomon Islanders reside in rural areas on customary land. Contemporary forms of 'community' in most areas are based on a complex interplay of kinship relations, exchange relations, friendships, church membership and myriad forms of claims to customary land of which genealogical descent is only one (see Keesing 1970; Hviding 1996; McDougall 2005; Scott 2007).

Despite calls from several 'federalist' islands and regions, particularly Guadalcanal and the western islands, for greater political devolution, a unitary system of Westminster government was introduced at the time of independence from Great

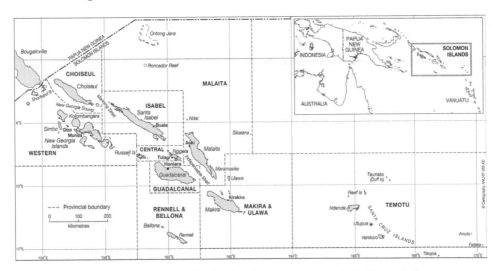

FIGURE 1. Solomon Islands, showing current provincial boundaries.

Britain in 1978 (Ghai 1983). However, the issues of devolution and provincial autonomy have continued to loom large in the politics of the independent state.

The ethnic tension

The initial phase of violence and disorder that occurred between late 1998 and July 2003 is referred to locally as the 'ethnic tension', or the *tenson* in Solomons *Pijin*. It saw the violent harassment of settlers in rural and peri-urban areas to the east and west of Honiara, most of whom originated from Malaita, by a militia of young Guale men initially calling themselves the Guadalcanal Revolutionary Army and, later, the Isatabu Freedom Movement (IFM). About 35 000 people, again mostly Malaitan, were displaced from their usual places of residence on Guadalcanal as a direct consequence of this violence. Most of these displacements took place between May and July 1999 and about 70 per cent of them occurred in rural wards in north Guadalcanal (Fraenkel 2004, p. 55; Moore 2004, p. 110).

The uprising commenced shortly after a speech made by the then Premier of Guadalcanal, Ezekiel Alebua, in which he put a number of demands to the national government. These were later reiterated in January 1999, in a submission signed by the members of the Guadalcanal Provincial Assembly titled 'Demands by the Bone Fide and Indigenous People of Guadalcanal'. The keynote demand was for state government for Guadalcanal under a federal system of government, a demand which had previously been put to the national government in 1988 in a document titled 'Petition by the Indigenous People of Guadalcanal'. Other 'Bone Fide' demands included: the return of alienated lands; the reform of land legislation to restrict ownership by people from other provinces; that Guadalcanal Province be granted 50 per cent of the revenue from resource projects on the island; and that legislation be introduced to 'control and manage' internal migration. (Fraenkel 2004, pp. 44–52, 197–203).

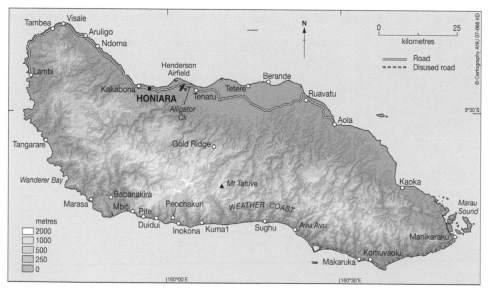

FIGURE 2. Guadalcanal.

From late 1999, a rival Malaitan militia, the Malaita Eagle Force (MEF), began to emerge and, after forming a 'joint operation' with para-military elements of the Malaitan-dominated Royal Solomon Islands Police Force, staged a *coup d'état* in June 2000. The spectre of all-out civil war was avoided with the signing of the Townsville Peace Agreement (TPA) in October 2000. However, varying degrees of violence and lawlessness continued to characterise the capital, Honiara, the Weather Coast of Guadalcanal and parts of Malaita and Western Province until the deployment of the Regional Assistance Mission to Solomon Islands (RAMSI) in July 2003.

Economic history, migration and identity politics

The establishment of the British Solomon Islands Protectorate in 1893 was premised on the requirement that the colony would be economically self-sufficient and this was to be achieved through the establishment of a plantation economy based on copra production (Bennett 1987, p. 103). For reasons of geography and resource disparities, the plantation economy reinforced economic divisions between Solomon Islanders—particularly between those in the east and those in the west—which had already started to develop over the preceding 40 years or so during the early trading period. This 'island based deprivation' (Bennett 1987, p. xviii) has driven labour migration and has been integral to the processes of island-wide and regional identity formation (see also Moore 2007).

Given their isolation from the early trading and colonial plantation economies, people from Malaita and the Weather Coast of Guadalcanal (as well as from other parts of the archipelago, notably Makira and Isabel) had no choice but to sell their labour on foreign shores, initially during the 'blackbirding' era, and then on the coconut plantations that were being developed in the western islands and on the north and east coasts of Guadalcanal from the first decade of the twentieth century (Corris 1973; Bennett 1987, p. 168). They received far fewer returns for their labour than those in the western islands and on northern Guadalcanal who were able to sell commodities, especially copra, directly to the European traders (Bennett 1987, p. 87).

The pre-Second World War pattern of what was effectively enclave development continued after the war and became increasingly institutionalised through the centralised planning process (Herlihy 1981, pp. 184–227). Significant disparities in village incomes were demonstrated in a number of detailed socio-economic studies conducted in the 1960s and 1970s (Lasaqa 1968; Chapman 1970; Bathgate *et al.* 1973; Bathgate 1973a, b, 1993; Herlihy 1981). The key distinction continued to be between north Guadalcanal and the western islands on the one hand, and the Weather Coast of Guadalcanal, Malaita and the eastern islands on the other (Bathgate 1993, p. 149). Spatial inequalities were especially stark on Guadalcanal where average per capita annual incomes were found to range from around AUS$20 at Nduindui and Pichahila (in 1965) on the Weather Coast, to around AUS$95 at Taboko (in 1971) on the northwest of the island (Bathgate 1993, table 9.12).

The perpetuation of pre-war patterns of spatial unevenness in village cash-earning opportunities, compounded by a decline in the delivery of education and health services, ensured that the phenomenon of labour migration also continued. Honiara and the adjacent northeast Plains of Guadalcanal, which became the focus of post-war commercial agricultural development, attracted migrants in increasing numbers (Bellam 1970). Prior to the *tenson*, the Solomon Islands Plantation Limited (SIPL) palm oil operation on the Guadalcanal Plains was employing about 1800 people, mostly Malaitans. When one includes their dependants, there were about 8000–10 000 migrant settlers associated with SIPL living on the Plains in the late 1990s (Fraenkel *et al.* 2010). The Gold Ridge mine in the mountains east of Honiara was another significant source of employment from the commencement of its construction in 1994 until its closure in June 2000.

As well as 'pull' motives in the form of employment and other economic and livelihood opportunities, land shortages caused by population growth and the expansion of cash-cropping have been an important 'push' motive driving the movement of people from Malaita, particularly from the densely populated north of the island (Frazer 1985a, p. 229; Gagahe 2000, p. 63; Bennett 2002, p. 13). The nature of Malaitan mobility has changed over time, with a trend towards permanent settlement as opposed to circulation, though both circulation and migration remain important (Kama 1979; Frazer 1985a, b; Gagahe 2000). Many of the settlers on Guadalcanal prior to the land evictions were in fact second- or third-generation residents (Connell 2006, p. 114).

Malaitan settlement: informal land transactions[2]

On the fertile Guadalcanal Plains east of Honiara settlement was taking place both on government-owned land via the Temporary Occupation License (TOL) system managed by the Lands Division, and on customary land (Kama 1979). In the case of the latter, migrants were able to obtain rights to live on and use customary land belonging to local matrilineal landowning groups known as *mamata*. Some of the early land use arrangements were documented by Lasaqa (1968) who conducted fieldwork in the Guadalcanal Plains area in 1966–67. At that stage, there were only a handful of Malaitans living on the Plains. Some of these men had obtained usufructuary rights through marriage into the *mamata*, but most were there at the invitation of local big-men, the heads of the *mamata* (Lasaqa 1968, pp. 162–6). Some settlers were selling sweet potato at the Honiara market, giving a share of the proceeds to the local *mamata* heads, while also working in the *mamata* heads' coconut groves; thereby paying 'rent' in the form of both cash and labour.

Kama, who conducted a survey of settlement on the Guadalcanal Plains in 1978–79, found a similar situation to that described a decade earlier by Lasaqa:

> Most of the people squatting on customary land made informal agreements with customary landowners, especially the head of the land holding group, through a friendly relationship. Some obtained permission through inter-marriage with members of the land holding groups. There had been no rents paid on land occupied, except in one case a cash payment for land was involved. (Kama 1979, p. 150; see also Frazer 1985b, pp. 196–7)

Recent accounts of land and settlement on Guadalcanal echo the earlier findings of Lasaqa and Kama (Maetala 2008; Solomon Islands Government 2010b). They continue to emphasise the central role of social relationships in securing and maintaining usufructuary rights to customary land. Such relationships are based both on the actions and behaviour of those coming into the landowning group, and on exchanges of traditional foodstuffs and wealth items, including cash, known as *chupu*. Importantly, the principles of Guale customary land tenure dictate that people from outside the matrilineal landowning clan can only ever be granted usufructuary rights, even when *chupu* has been paid, and that these rights are not automatically passed on at death.

One perspective on the role of land and settlement in the *tenson* is that, over time, as migrant men built houses and made gardens, and were joined by their families and *wantoks*, the land boundaries and sheer number of 'tenants' extended 'beyond those originally agreed between the first settler and his vendor or landlord' (Bennett 2002, p. 8). The situation was compounded by the fact the workers at SIPL, and their families, were building houses and making gardens on customary land adjacent to the designated staff housing areas (UNDP 2004, p. 26). This sort of overspill onto customary land was also taking place around some of the TOL areas.

The sources examined here encourage us to consider two further factors that contributed to the land evictions: the breakdown of social relationships between settlers and landowning groups; and the breakdown in relationships amongst landowners themselves. These factors share a salient generational dimension. In the case of the former, Maetala emphasises the role of 'good deeds', 'actions' and 'behaviour' in underpinning outsiders' rights to customary land on Guadalcanal (2008, pp. 46–7). Making traditional payments will not, in itself, guarantee the transfer of these rights to an outsider's family upon his death: 'to be entitled to pass down that land, his actions and behaviour during his stay with the landowning group must be seen to be "good"' (Maetala 2008, p. 46).

Importantly, the stories of these good deeds are increasingly forgotten as 'generations pass without the retelling of tribal history', which can give rise to conflict (Maetala 2008, p. 46). In the case of Malaitan settlers, Maetala notes that relationships have frequently broken down because of the failure on the part of the *wantoks* who have come to live with them 'to demonstrate the same respect shown by their kin for their hosts' (2008, p. 47). This issue of cultural respect is frequently raised by Guales when discussing the 'causes' of the *tenson* (see Liloqula & Pollard 2000, p. 6; Allen 2007, pp. 143–5).

Culture has also played a central role in conflicting interpretations of customary land tenure on Guadalcanal. The recent consultations of the Commission of Inquiry into Land Dealings and Abandoned Properties on Guadalcanal have underscored the widely held view amongst Guales that outsiders can, upon payment of *chupu*, obtain rights to *use* customary land, but never to *own* it outright. Some Guale landowners believe that this clashes with Malaitan understandings of transacting customary land. For example, during the Commission's consultation in the Giana region west of Honiara a female landowner stated:

> Our problem is when people from other provinces come in, they have different ways of dealing with land. Their land tenure is different from ours and therefore when we give them land and they give us *chupu* or they give us cash, perhaps to them, they are buying the land as they buy something from the market or from the shop, which is not our custom traditionally ... (Solomon Islands Government 2010a, p. 10)

Participants in the Giana consultation also spoke about the problems and disputes created within landowning groups as a result of individual men selling customary land for cash to outsiders without the knowledge or permission of the other clan members. The commodification of customary land has 'created an environment ripe for enmity among families and clan members' (Maetala 2008, p. 39). The relationship between land sales, on the one hand, and inter-group and intergenerational conflict, on the other, is described by Kabutaulaka as follows:

> Many individuals were selling land without consulting other members of their line (*laen*, tribe) [*sic*], often causing arguments among landowners ... Over the years the sale of land has been resented by a younger generation of Guadalcanal people who view it as a sale of their birthright. (2002, p. 7)

Guale discontent

Since the 1950s, Guales from both the north and south of the island have repeatedly expressed their concerns about the increasing numbers of migrants present in and around Honiara and on the Guadalcanal Plains. In 1954, the Special Lands Commissioner, whilst touring northeast Guadalcanal, noted that 'the worst fear the Tasimboko people have is in regard to the immigration of Malaita people' (Allan 1989, p. 79). He also found similar concerns on the Weather Coast and at Marau (Allan 1989, p. 87). In 1980, MP for Central Guadalcanal (which includes part of the coast east of Honiara) Paul Tovua called on the national parliament to control the movement of people between provinces (Alasia 1989, p. 119). Tovua was concerned that 'certain of the migrants [in his constituency] did not show respect for their hosts or for their hosts' culture' (Alasia 1989, p. 119).

Freedom of movement had been an issue discussed in the constitutional debates leading up to independence, and grievances regarding internal migration characterised the submissions to the Constitutional Review Committee of 1987 (Chapman 1992). Concerns were voiced in relation to migrants settling on land without the 'permission', or 'proper consent', of the customary land owners, and 'to the impact of newcomers on the integrity of long-established customs' (Chapman 1992, p. 94). These submissions were accompanied by demands that a federal system of government be introduced.

It is noteworthy that some of these concerns were being articulated by people from the Weather Coast region. Large numbers of men from south Guadalcanal went to work in and around Honiara and on the Guadalcanal Plains after the war, and these movements have become increasing permanent over the past 30 years or so, making settlers from south Guadalcanal important 'actors' in the issues surrounding migration and land on north Guadalcanal (Bennett 1987, p. 314; 2002, p. 12). In their submission to the Constitutional Review Committee, the Moro Movement[3] questioned whether an individual's right to freedom of movement should be absolute: 'People from other islands should have the right to travel from one place to another [but] they ... should not have the right to reside or settle [in other places] without permission of those in authority' (cited in Chapman 1992, p. 92).

Frustrations surfaced again in 1988 when Guadalcanal people demonstrated following multiple murders near Honiara (see Keesing 1992, pp. 178–9). In their petition to the national government (referred to as the 'Petition by the Indigenous

People of Guadalcanal'), they requested the establishment of a federal system of government and that 'immediate steps be taken to reduce the pressure of internal migration' (quoted in Fraenkel 2004, p. 196). These demands were largely ignored by the national government led by Ezekiel Alebua.

Competing narratives of development: indigenous people and nation builders

Guadalcanal: the contested motherland

The texts of my interviews with Guale ex-militants and other members of the Guale community are peppered with the words 'not fair', 'unfair', 'resources', 'the government' and 'development'. They believe that it is unfair that their province provides a significant amount of revenue to the national economy from resource developments but receives proportionately much less from the national government in terms of grants and disbursements. The inadequate state of infrastructure and service provision on Guadalcanal is directly linked to this state of affairs and the blame is laid squarely on 'the government'.

An important aspect of this grievance, which I label 'development equity', is the perception amongst Guales that their land and resources are being used to develop and benefit 'other' people, whilst Guadalcanal and its people are being neglected and forgotten. These sentiments are expressed particularly strongly by Weather Coast people, and it is important to bear in mind that all of the Guale militant leaders originally hail from the Weather Coast region.

The following statements, the first three from Weather Coast ex-militants and the fourth from a Birao-speaking (i.e. Guale) chief from the Marau area, exemplify this perception of others benefiting from Guale land and resources:

> We oppose development in terms of ripping it out and building up different people, different nations, different places. (Interview with M)

> All the resources which they harvest from Guadalcanal they use for different people, from other provinces. (Interview with K)

> I wanted a country where my people's resources are used for their benefit and not that of 'strangers'. (Gray 2002, p. 6)

> More benefits should be coming to people on Guadalcanal, for developing the rural areas or parts of town, but that's not happening. They are neglected and other people benefit. (Interview with R)

The development equity grievance is often explicitly linked to the desire for 'statehood' for Guadalcanal. From the perspective of many Guales, the primary advantage of greater autonomy is that they will receive a much larger proportion of the economic benefits flowing from the resource developments which take place on their island and have a much greater say in how those resources are developed. According to the ex-militant K: 'Mostly state government is what we really want so that we have power over our own resources.' Similarly, J states: 'If it [the government] addresses this issue [state government], everyone will go back to their places and Guadalcanal will become one of the richest islands ... we have all the resources' (interview with J).

As we have seen, Guale claims to land and resources have been framed by claims of indigeneity, as in the 1998 'Demands by the Bone Fide and Indigenous People of Guadalcanal' and the 1988 'Petition by the Indigenous People of Guadalcanal'. During the *tenson*, Guale militants sought to demonstrate, or perform, indigeneity in a number of ways, including by wearing the *kabilato*, the traditional dress of the Gaena'alu Movement (formerly known as the Moro Movement); by claiming 'Isatabu' as a pre-contact name for the island of Guadalcanal; and by invoking *kastom* and ancestral connections to land (see Kabutaulaka 2001). Militant leader Andrew Te'e, for example, published a series of three articles titled 'Land is sacred to me' in which he discusses the need to uphold the ancient 'Divine Law' which governs the custodianship of 'mama Isatabu' (Te'e 2000).

The view from Malaita: building the nation, saving the nation

In contrast to the Guale narrative of the exploitation of their land and resources at the hands of successive waves of outsiders, Malaitans bemoan the deliberate neglect and under-development of their island by colonial and postcolonial governments. Rather than invoking indigeneity as the basis for establishing legitimate rights to land and access to economic opportunities, Malaitans cast themselves as the workers and builders of the nation, thereby linking themselves to the legitimacy of the state and its broader modernising project, in much the same way migrants have done in West New Britain Province of Papua New Guinea (Koczberski & Curry 2004).

Interviews with ex-militants, as well as a number of recently published sources, again provide informative windows onto this identity narrative. Many Malaitan ex-militants with whom I spoke describe their actions and motivations during the *tenson* in terms of 'saving the nation'. This objective, to save the nation, can be located within a deeper historical narrative of Malaitans as nation builders: having built the nation it was incumbent upon them to rescue it from an incompetent government, a hamstrung police force and the Guale militants.

The following is an exemplar articulation, by a former Malaitan militant, of the belief that the Malaita Eagle Force saved Solomon Islands:

> Due to what we did, we restored Solomon Islands ... We saved not only Malaita, but the whole Solomon Islands. Due to what we did, we saved all Solomon Islands. (Interview with T)

Malaitan ex-militants also frequently spoke about the historical role of Malaitans as the 'manpower' of Solomon Islands. For example:

> You can only call Solomon Islands a country because of Malaitans. Without the hands of Malaitans you would not see any development like palm oil. There wouldn't be anyone to provide labour for these things. (C, interview with C, L and others)

These expressions of a Malaitan 'story' are, not surprisingly, shared by others in the wider Malaitan community. They are articulated by the Malaita Ma'asina Forum, a non-government organisation which formed in September 2003 'as a voice to raise concerns and issues affecting Malaita and the people of Malaita'

(Malaita Ma'asina Forum Executive 2005, p. 3). While the Ma'asina Forum evokes obvious connotations with the anti-colonial Maasina Rule Movement of the late 1940s and early 1950s, its membership base is unknown and it should not be seen as representative of the views of a majority of Malaitans. That said, the Forum provides an exemplar statement of the Malaitan view of the conflict and of the role of Malaitans in the nation:

> ... the MEF was ... protecting Honiara city which was under serious threat from the Guadalcanal rebels ... Malaitan manpower has contributed greatly towards the development of this whole nation. (Malaita Ma'asina Forum Executive 2005, p. 16)

In a booklet examining the *tenson*, titled *Trouble in paradise*, former national parliamentarian the Revd Michael Maeliau describes the 'Guadalcanal Cause' and the 'Malaita Case', attempting in both cases to locate these stories in the 'objective' history of Solomon Islands (Maeliau 2003). Like the ex-militants cited above, Maeliau portrays Malaitans as the builders of the Solomon Islands: 'the fact remains that with their blood, sweat and tears they contributed to laying the foundation and the building up of this nation' (2003, p. 51).

These themes are also developed by Michael Kwa'ioloa, a Malaitan (Kwara'ae) chief and former policeman, in his published account of the origins and dynamics of the conflict (Kwa'ioloa & Burt 2007, pp. 114–15). Kwa'ioloa describes the role of Malaitans in the development of Guadalcanal and Honiara, contrasting the industriousness and productivity of Malaitans against those who expect 'things to come to them without work and sweat' (Kwa'ioloa & Burt 2007, p. 114). He states that it was through the efforts of Malaitans in the Second World War that the Americans were able to win the battle of Guadalcanal; that Malaitans cleared and planted the plantations east and west of Honiara; that they built Honiara itself; and, most recently, that they provided the workforce for Gold Ridge mine, 'operating the machines and earning revenue for the government and royalties for Guadalcanal people from their land' (Kwa'ioloa & Burt 2007, p. 115).

The sentiment that the MEF saved not only Malaitans and other people in Honiara but the whole of Solomon Islands can, then, be understood in this broader narrative of Malaitan identity: having built the nation (and saved it once already during the Second World War), it was incumbent upon Malaitans to protect it and ultimately to save it from the Guale militants, an incompetent national government and an ineffectual police force.

Conclusion: projects of exclusion or projects of resistance?

We have seen that social relationships have played a central role in the story of settlement and eviction on Guadalcanal. The establishment of relationships enabled Malaitan settlers to gain and maintain usufructuary rights to customary land. These relationships often started out as a simple friendship, were cemented by gift exchange, and then maintained over time, even over generations, by the 'good deeds' of the initial settler and their retelling in 'tribal history'. The disintegration of these relationships occurred when these stories were no longer told or when newly arrived *wantoks* ceased to perform the good deeds of the original settler and failed to respect the culture of their hosts. Misunderstandings also took place around conflicting interpretations of transacting customary land (cf. Koczberski *et al.* 2009, p. 34).

It seems that the breakdown of relationships within landowning groups was just as pertinent to the land evictions as the disintegration of relationships between settlers and landowners. These internecine disputes occurred when senior men within Guale landowning groups effectively sold customary land to settlers without the knowledge or approval of the other members of their groups. Such disputes also had a salient intergenerational dimension as young Guale men came to resent the sale of 'their birthright'.

The centrality of social relationships in establishing and maintaining settlers' rights to customary land has been recently demonstrated by Curry and Koczberski (2009) and Koczberski *et al.* (2009) in the case of the oil palm 'frontier regions' of West New Britain and Oro Provinces in Papua New Guinea. There are striking similarities with the situation on Guadalcanal. In the case of West New Britain, those settlers who accept the social embeddedness of land transactions and act 'in accordance with traditional rules and expectations' have generally enjoyed stable relationships with their host communities and, therefore, 'ongoing access to land' (Koczberski *et al.* 2009, p. 38).

However, Koczberski and colleagues also make the point that not all settlers have been able to achieve security of access under this framework of 'property rights for social inclusion'. There are many instances of settlers losing their access rights and, with direct similarities to the Guadalcanal case, 'being evicted or harassed by members of the landowning group, especially by younger clan members' (Koczberski *et al.* 2009, p. 34).

Yet West New Britain has never experienced mass evictions or 'ethnic tensions' on the scale of those that took place on Guadalcanal. This is an interesting question to ponder. I suggest that the main difference between the two cases is that localised grievances around land and settlement on north Guadalcanal have, for a relatively long time, been mobilised to the larger project of greater autonomy for Guadalcanal. This project has involved a much wider set of grievances, most notably around the distribution of development benefits and government services. Moreover, it has involved a broader range of actors. Weather Coast people, who have long been marginalised from the enclave developments on the northern side of the island, have actively sought to press their claims to 'development' by limiting those of 'outsiders', including the state.[4] Bainton describes a similar process occurring on Lihir Island in Papua New Guinea, where the majority of Lihirians, who do not receive royalties or compensation from the mine, have reassessed their relationships with non-Lihirians. Previously accepted as visitors, non-Lihirians are now viewed as strangers attempting to access benefits associated with the mine; as an 'encroachment ... on *their* development' (Bainton 2009, p. 23; original emphasis).

Compensation for lost land and properties on Guadalcanal became a critical concern for Malaitans following the land evictions of 1999. This was reflected in Part Four of the TPA which committed the Solomon Islands government to establishing a commission of inquiry to investigate the acquisition of land on Guadalcanal by non-Guadalcanal persons prior to 1998. The Commission, referred to above, was finally established in late 2009. However, it appears that few claimants have stepped forward, perhaps reflecting the widely held view amongst Malaitans that they no longer want to live and work in rural Guadalcanal (see, for example, Maeliau 2003, p. 52).

Indeed, the events of the *tenson* appear to have precipitated a revival of Malaitan introspection strongly reminiscent of the Maasina Rule period. This is particularly apparent in the work and objectives of the Malaita Ma'asina Forum which is attempting to facilitate the development of Malaita by lobbying the government for infrastructure and investment in agriculture projects, and encouraging the recording and registering of customary land (Malaita Ma'asina Forum Executive 2005, p. 17). These objectives are shared by many Malaitans with whom I have spoken over the past 6 years (also see Moore 2007, pp. 227–31). They represent a renewed attempt to create the economic incentives for Malaitans to stay on Malaita, just as Maasina Rule had attempted to do after the Second World War (see Keesing 1982; Laracy 1983).

While Malaitans may have become ambivalent about harnessing state law to press for the reinstatement of their land and property rights on rural Guadalcanal (beyond securing compensation), their identity narrative nevertheless remains inextricably tied to the legitimacy of the state. The state underpins the rights of Malaitans to live within the Honiara town boundary—where they continue to comprise a significant proportion both of the urbanised elite and the town's overall population—and also on government-owned land surrounding Honiara. Moreover, the objective of developing Malaita can only be realised within the current frame-work of a unitary Solomon Islands where the central government retains the power to make decisions about infrastructure investment, developing planning, and the redistribution of resource rents.

Guales, by contrast, continue to challenge the legitimacy of the state in matters of land and the management of the revenues that flow from the exploitation of natural resources on Guadalcanal. They remain committed to 'statehood' for Guadalcanal to be realised through the implementation of a new federal constitution, a draft of which has been prepared in accordance with Part Four of the TPA. The resolutions of the 2005 Guadalcanal Leaders Summit, which involved 300 government and civil society representatives from throughout Guadalcanal, reiterated many of the 1999 'Bone Fide' demands, making extensive reference to land and resource issues, and the adoption of the proposed federal constitution (Anon. 2005).

In the wake of the *tenson*, the need for labour-intensive development projects on Malaita is more pressing than ever. As well as their strong sense of no longer being welcome in rural Guadalcanal, there are new structural barriers preventing Malaitans from working on the commercial agricultural projects for which they have historically provided the bulk of the workforce. For example, Guadalcanal Plains Palm Oil Limited, which took over the old SIPL plantations in 2004, has signed a memorandum of understanding with landowners and the provincial and national governments which dictates that first priority with regard to employment is to be given to local landowners, then to people from elsewhere in Guadalcanal Province and, finally, to people from other provinces (Fraenkel *et al.* 2010). Similar arrangements are also in place at Gold Ridge mine which recommenced production in early 2011 (Nanau 2009, pp. 193, 251).

However, developing Malaita will be no easy task, as discussed recently by Moore (2007, pp. 227–31). It will require land reforms which are carefully attuned to the complexities of the island's myriad land systems and an approach to develop-ment which is strongly informed by Malaitan epistemologies of rural development (see also Gegeo 1998). Without this sort of accommodation with local visions for development and forms of land tenure, it will be extremely difficult to effect large-scale infrastructure and agricultural development on Malaita (Moore 2007, pp. 227–31).

Indeed, the same may be said of 'development' on Guadalcanal, in parts of Western Province (Hviding 1993, 2003; Scales 2004) and, in fact, throughout much of Melanesia. While the central focus of this paper has been on land and settlement on Guadalcanal, we have also seen that a critical issue for many Guales has been the nature of development itself. In 2006 the current leader of the Gaena'alu Movement told me that the Movement is opposed to developments like Gold Ridge mine. Such developments are at odds with the Movement's philosophy of preserving and protecting the land, resources and people of Guadalcanal. To use his words, they are 'too big' (interview with Jerry Sabino 19 April 2006). From this perspective, Guale calls for greater autonomy are better seen in terms of a desire to resist the imposition of the state in matters of land, resources and development, as opposed to a desire to capture and maximise the economic benefits flowing from the resource wealth of their island.

There is increasing recognition that Melanesians' engagements with capitalist economies are inflected to serve place-based socio-economic and cultural goals as they seek to achieve a meaningful blend of modernity and tradition (Curry 2003). For the Kwara'ae speakers of Malaita, this is expressed as *gwaumauri'anga* or 'good life' (Gegeo 1998). This resonates with the Papua New Guinea *Tok Pisin* term *gutpela sindaon*, again referring to a localised conception of the good life. These conceptions underscore the salience of local epistemologies of development, of the dialectics of people and nature, and of the connections between land and identity. As has been the case in other parts of Melanesia (see Banks 2008) and in other developing-country contexts (see, for example, Peluso & Watts 2001), conflict can result from the disruption to these relationships as local socio-ecological contexts are increasingly penetrated by the resource-hungry global economy.

Acknowledgements

I am grateful to two of the editors of this special issue, Gina Koczberski and George Curry, for reading an earlier draft of this paper and making very useful suggestions for revisions. I also received some useful comments from two anonymous referees. I take full responsibility for any errors of fact or interpretation.

NOTES

[1]   The interviews were conducted in the context of a research project examining the motives and experiences of men who joined the rival militant groups during the conflict of 1998–2003 (Allen 2007). As such the voices examined are primarily those of former militants. However, some material from interviews with non-combatants is also presented, and the perspectives of traditional leaders, politicians, intellectuals and

local non-government organisations are gleaned from published sources. Most of the interviews were conducted during a 9-month period of fieldwork on Guadalcanal and North Malaita in 2005 and 2006.

[2]   This section focuses on settlement in the rural wards east of Honiara, which is where most of the displacements and evictions occurred during the *tenson*. That said, many of the issues discussed equally apply to the settlements that started to emerge around Honiara in the 1970s (see Alasia 1989; Storey 2003; Chand & Yala 2008) and to the peri-urban areas west of Honiara (Monson 2010).

[3]   The Moro Movement, now known as the Gaena'alu Movement, is a 'back to custom', anti-colonial movement that emerged on the Weather Coast in the late 1950s (Davenport & Çoker 1967; Bennett 1987, pp. 316–17). While the precise extent of the Movement's current influence and membership is not known, its leadership claims several thousand followers mostly on the eastern side of the Weather Coast and adjacent hinterland.

[4]   Along with 'Guadalcanal' and 'Malaita', the third large island- or regional-wide ethnic grouping in Solomon Islands is 'Western' or 'West', referring to Western Province and sometimes including Choiseul Province. Like Guadalcanal, Western is relatively resource rich as a result of its timber resources, and revenue-sharing arrangements with the central government were a central concern of the 'Western Breakaway Movement' that emerged on the eve of independence in 1978. Western has not experienced in-migration and settlement to the same degree as Guadalcanal, and nor does it share the latter's sharp internal socio-economic divisions. That said, the Breakaway Movement was driven, in part, by anxieties about becoming part of an independent nation dominated by the 'Malaitan Mafia' (Bennett 1987, pp. 327–30).

REFERENCES

ALASIA, S. (1989) 'Population movement', in Laracy, H. & Alasia, S. (eds) *Ples Blong Iumi: Solomon Islands, the past four thousand years*, Institute of Pacific Studies, Suva.

ALLAN, C.H. (1989) *Solomons safari 1953–58* (Part 1), Nag's Head Press, Christchurch.

ALLEN, M.G. (2007) 'Greed and grievance in the conflict in Solomon Islands, 1998–2003', unpublished PhD thesis, Australian National University, Canberra.

ANON. (2005) 'Guadalcanal Leaders Summit Resolutions', unpublished document, 18 February.

BAINTON, N.A. (2009) 'Keeping the network out of view: mining, distinctions and exclusion in Melanesia', *Oceania* 79(1), pp. 18–33.

BANKS, G. (2008) 'Understanding "resource conflicts" in Papua New Guinea', *Asia Pacific Viewpoint* 49(1), pp. 23–34.

BATHGATE, M.A. (1973a) 'A study of economic change and development in the indigenous sector, West Guadalcanal, British Solomon Islands Protectorate', Victoria University, Wellington.

BATHGATE, M.A. (1973b) 'Bihu matena golo: a study of the Ndi-Nggai of West Guadalcanal and their involvement in the Solomon Islands cash economy', PhD thesis, Victoria University of Wellington.

BATHGATE, M.A. (1993) *Fight for the dollar: economic and social change in western Guadalcanal, Solomon Islands*, Alexander Enterprise, Wellington.

BATHGATE, M.A., FRAZER, I. & McKINNON, J. (1973) 'Socio-economic change in Solomon Islands villagers: summary team report of the Victoria University of Wellington socio-economic study of the B.S.I.P.', Victoria University, Wellington.

BELLAM, M. (1970) 'The colonial city: Honiara, a Pacific Islands case study', *Pacific Viewpoint* 11(1), pp. 66–96.

BENNETT, J.A. (1987) *Wealth of the Solomons: a history of a Pacific archipelago 1800–1978*, University of Hawaii Press, Honolulu.

BENNETT, J.A. (2002) 'Roots of conflict in the Solomon Islands: though much is taken, much abides: legacies of tradition and colonialism', State, Society and Governance in Melanesia Project, Discussion Paper 2002/5, Research School of Pacific and Asian Studies, Australian National University, Canberra.

CHAND, S. & YALA, C. (2008) 'Informal land systems within urban settlements in Honiara and Port Moresby', in *Making land work* (Volume 2), Commonwealth of Australia, Canberra, pp. 85–105.

CHAPMAN, M. (1970) 'Population movement in tribal society: the case of Duidui and Pichahila, British Solomon Islands', PhD thesis, University of Washington.

CHAPMAN, M. (1992) 'Population movement: free or constrained? The first 10 years of Solomon Islands independence', in Crocome, R. & Tuza, E. (eds) *Independence, dependence, interdependence*, Institute of Pacific Studies, University of the South Pacific Honiara Centre and the Solomon Islands College of Higher Education, Suva and Honiara, pp. 75–97.

CONNELL, J. (2006) '"Saving the Solomons": a new geopolitics in the "arc of instability"?', *Geographical Research* 44(2), pp. 111–22.

CORRIS, P. (1973) *Passage, port and plantation: a history of Solomon Islands labour migration 1870–1914*, Melbourne University Press, Melbourne.

CURRY, G. (2003) 'Moving Beyond Postdevelopment: Facilitating Indigenous Alternatives for "Development"', *Economic Geography* 79(4), pp. 405–23.

CURRY, G. & KOCZBERSKI, G. (2009) 'Finding common ground: relational concepts of land tenure and economy in the oil palm frontier of Papua New Guinea', *Geographical Journal* 175(2), pp. 98–111.

DAVENPORT, W.H. & ÇOKER, C. (1967) 'The Moro Movement of Guadalcanal, British Solomon Islands Protectorate', *Journal of the Polynesian Society* 76, pp. 123–75.

ELMHIRST, R. (2001) 'Resource struggles and the politics of place in North Lampung, Indonesia', *Singapore Journal of Tropical Geography* 22(3), pp. 284–306.

ESCOBAR, A. (2001) 'Culture sits in places: reflections on globalism and subaltern strategies of localization', *Political Geography* 20, pp. 139–74.

FILER, C. (1997) 'Compensation, rent and power in Papua New Guinea', in Toft, S. (ed.) *Compensation for resource development in Papua New Guinea*, Australian National University, Canberra, pp. 156–89.

FRAENKEL, J. (2004) *The manipulation of custom: from uprising to intervention in the Solomon Islands*, Pandanus Books, Canberra.

FRAENKEL, J., ALLEN, M.G. & BROCK, H. (2010) 'The resumption of palm-oil production on Guadalcanal's northern plains', *Pacific Economic Bulletin* 25(1), pp. 64–75.

FRAZER, I. (1985a) 'Circulation and the growth of urban employment amongst the To'ambaita, Solomon Islands', in Chapman, M. & Prothero, R.M. (eds) *Circulation in population movement: substance and concepts from the Melanesian case*, Routledge & Kegan Paul, London.

FRAZER, I. (1985b) 'Walkabout and urban movement: a Melanesian case study', *Pacific Viewpoint* 26(1), pp. 185–205.

GAGAHE, N.K. (2000) 'The process of internal movement in Solomon Islands: the case of Malaita 1978–1986', *Asia Pacific Population Journal* 15(2), pp. 53–75.

GEGEO, D.W. (1998) 'Indigenous knowledge and empowerment: rural development examined from within', *The Contemporary Pacific* 10(2), pp. 289–315.

GHAI, Y. (1983) 'The making of the independence constitution', in Larmour, P. & Tarua, S. (eds) *Solomon Islands Politics*, Institute of Pacific Studies, Suva, pp. 9–52.

GRAY, G. (2002) 'Habuna Momoruqu (The Blood of my Island): violence and the Guadalcanal uprising in Solomon Islands', paper presented at the Australian Anthropological Society annual conference 'Anthropology and Diversity', 3–5 October, Australian National University, Canberra.

HERLIHY, J.M. (1981) 'Always we are last: a study of planning, development and disadvantage in Melanesia', PhD thesis, Australian National University, Canberra.

HVIDING, E. (1993) 'Indigenous essentialism? Simplifying customary land ownership in New Georgia, Solomon Islands', *Bijdragen tot de Taal-, Land- en Volkenkunde* 149(4), pp. 802–24.

HVIDING, E. (1996) *Guardians of the Marovo Lagoon: practice, place and politics in maritime Melanesia*, University of Hawaii Press, Honolulu.

HVIDING, E. (2003) 'Contested rainforests, NGOs, and projects of desire in Solomon Islands', *International Social Science Journal* 55(4), pp. 539–54.

KABUTAULAKA, T. (2001) 'Beyond ethnicity: the political economy of the Guadalcanal crisis in the Solomon Islands', Working Paper 01/01, State, Society and Governance in

Melanesia Project, Research School of Pacific and Asian Studies, Australian National University, Canberra.

KABUTAULAKA, T. (2002) 'A weak state and the Solomon Islands peace process', East–West Centre Working Paper No. 14, Centre for Pacific Island Studies, University of Hawaii, Manoa.

KAMA, T. (1979) 'Guadalcanal Plains', in Heath, I. (ed.) *Land research in Solomon Islands*, Land Research Project, Lands Division, Ministry of Agriculture and Lands, Honiara.

KEESING, R.M. (1970) 'Shrines, ancestors, and cognatic descent: the Kwaio and Tallensi', *American Anthropologist* 72(4), pp. 755–75.

KEESING, R.M. (1982) 'Kastom and anticolonialism on Malaita: "culture" as political symbol', *Mankind* 13(4), pp. 357–73.

KEESING, R.M. (1992) *Custom and confrontation: the Kwaio struggle for cultural autonomy*, University of Chicago Press, Chicago and London.

KOCZBERSKI, G. & CURRY, G.N. (2004) 'Divided communities and contested landscapes: mobility, development and shifting identities in migrant destination sites in Papua New Guinea', *Asia Pacific Viewpoint* 45(3), pp. 357–71.

KOCZBERSKI, G., CURRY, G.N. & IMBUN, B. (2009) 'Property rights for social inclusion: migrant strategies for securing land and livelihoods in Papua New Guinea', *Asia Pacific Viewpoint* 50(1), pp. 29–42.

KWA'IOLOA, M. & BURT, B. (2007) 'The Chiefs' Country': A Malaitan View of the Conflict in Solomon Islands', *Oceania* 77(1), pp. 111–27.

LARACY, H. (ed.) (1983) *Pacific protest: the Maasina Rule Movement, Solomon Islands, 1944–1952*, Institute of Pacific Studies, University of the South Pacific, Suva.

LASAQA, I.Q. (1968) 'Melanesians' choice: a geographical study of Tasimboko participation in the cash economy, Guadalcanal, British Solomon Islands', PhD thesis, Australian National University, Canberra.

LI, T.M. (2000) 'Articulating indigenous identity in Indonesia: resource politics and the tribal slot', *Comparative Studies in Society and History* 42(1), pp. 149–79.

LILOQULA, R. & POLLARD, A.A. (2000) 'Understanding conflict in Solomon Islands: a practical means to peacemaking', Discussion Paper 00/7, State, Society and Governance in Melanesia Project, Australian National University, Canberra.

MAELIAU, M. (2003) *Trouble in paradise*, Aroma Ministries, Honiara.

MAETALA, R. (2008) 'Matrilineal land tenure systems in Solomon Islands: the cases of Guadalcanal, Makira and Isabel Provinces', in Huffer, E. (ed.) *Land and women: the matrilineal factor*, Pacific Islands Forum Secretariat, Suva, pp. 35–72.

MALAITA MA'ASINA FORUM EXECUTIVE (2005) *Building peace and political stability in Solomon Islands: Malaita Ma'asina Forum perspective*, Solomon Islands Publications and Information Distribution Centre, Honiara.

MCDOUGALL, D. (2005) 'The unintended consequences of clarification: development, disputing and the dynamics of community in Ranongga, Solomon Islands', *Ethnohistory* 52(1), pp. 81–109.

MONSON, R. (2010) 'Women, state law and land in peri-urban settlements on Guadalcanal, Solomon Islands', Justice for the Poor Briefing Note 4(3), April.

MOORE, C. (2004) *Happy isles in crisis*, Asia Pacific Press, Canberra.

MOORE, C. (2007) 'The misappropriation of Malaitan labour', *Journal of Pacific History* 42(2), pp. 211–32.

NANAU, G. (2009) 'Can a theory of insecure globalization provide better explanations for instability in the South Pacific? The case of Solomon Islands', PhD thesis, University of East Anglia, Norwich.

PELUSO, N.L. & WATTS, M. (2001) 'Violent environments', in Peluso, N.L. & Watts, M. (eds) *Violent environments*, Cornell University Press, Ithaca, NY and London, pp. 3–38.

SCALES, I. (2003) 'The social forest: landowners, development conflict and the state in Solomon Islands', PhD thesis, Australian National University, Canberra.

SCOTT, M.W. (2007) *The severed snake: matrilineages, making place and a Melanesian Christianity in southeast Solomon Islands*, Carolina Academic Press, Durham, NC.

SOLOMON ISLANDS GOVERNMENT (2010a) Commission of Inquiry into Land Dealings and Abandoned Properties on Guadalcanal. Consultation on Bolomona Customary Land

Issues. Transcript of preliminary investigation into terminology 'tribe', 'clan', 'line', available from: http://www.comofinquiry.gov.sb (accessed 12 August 2010).

SOLOMON ISLANDS GOVERNMENT (2010b) Commission of Inquiry into Land Dealings and Abandoned Properties on Guadalcanal. Consultation on Giana Customary Land Issues, available from: http://www.comofinquiry.gov.sb (accessed 12 August 2010).

STOREY, D. (2003) 'The peri-urban Pacific: from exclusive to inclusive cities', *Asia Pacific Viewpoint* 44(3), pp. 259–79.

TE'E, A. (2000) 'Land is sacred to me', *Isatabu Tavuli* 1, pp. 1–3.

UNITED NATIONS DEVELOPMENT PROGRAM (UNDP) (2004) *Solomon Islands peace and conflict development analysis: emerging priorities in preventing future conflict*, United Nations Development Program, Honiara.

Changing Land Tenure and Informal Land Markets in the Oil Palm Frontier Regions of Papua New Guinea: the challenge for land reform

GINA KOCZBERSKI, GEORGE N. CURRY & JESSE ANJEN, *Curtin University, Australia; Curtin University, Australia; PNG Oil Palm Research Association, Papua New Guinea*

ABSTRACT *This paper reports on the authors' ongoing research with agricultural extension services, customary landowners and migrant farmers to develop a template for a Land Usage Agreement (LUA) that seeks to reconcile customary landowners' and migrants' differing interpretations of the moral basis of land rights. The LUA shows a way forward for land reform that builds on customary tenure while strengthening the temporary use rights of migrants to enable them to generate viable and relatively secure livelihoods. The paper concludes that land tenure reform should draw on what is already happening on the ground, rather than impose external models that do not accord with local cultural mores about the inalienability of customary land and its enduring social and cultural significance for customary landowning groups.*

Introduction

This paper examines informal land transactions between customary landowners (hereafter, landowners) and migrants seeking customary land for small-scale oil palm development in the provinces of West New Britain (WNB) and Oro, Papua New Guinea (PNG) (see Figure 1). The land transfers documented in this paper are occurring in an environment where the demand for land by 'outsiders' is high, where customary tenure is undergoing rapid change and where an effective land administrative system is absent (see Fingleton 2004). The pressure on customary land in PNG emanates from the long-term trend of people migrating from resource-poor areas to urban and peri-urban areas and to rural regions of the country offering relatively good access to employment, education and health services (e.g. mine sites and plantations) (see Numbasa & Koczberski this issue). With land reform back on the policy agenda in PNG, and in other parts of the

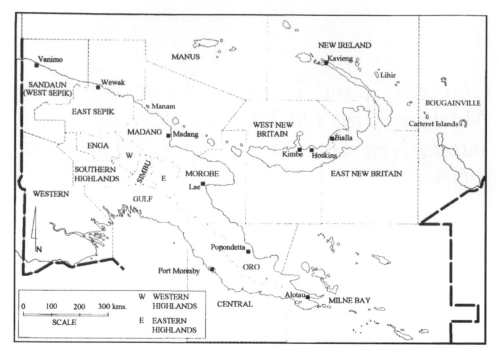

FIGURE 1. Location of study sites.

developing world, it is timely to consider what lessons can be learned from these informal land transactions to inform institutional approaches to land reform that build on existing tenure systems and are tailored more closely to local circumstances.

Landowners in PNG are not alone in responding to the large demand for land by 'outsiders' through informally 'selling', 'leasing' or gifting land. Across many parts of the Global South where land remains under customary tenure there is a proliferation of informal (and often illegal) land transfers occurring in rural and peri-urban areas as land-poor migrants enter into agreements with customary landowners to gain short- and long-term access to land (e.g. Amanor & Diderutuah 2001; Elmhirst 2001; Li 2002; Sjaastad 2003; Potter & Badcock 2004; Daley 2005; Chimhowu & Woodhouse 2006; Hagberg 2006; Martin 2007; Peters 2007; Koczberski *et al.* 2009; Chand & Yala 2012). Whilst informal land transactions vary greatly in type, typically most involve cash payments for the sale or leasing of customary land to outsiders and some form of written documentation.

Although the large demand and pressures on customary land by 'outsiders' helps explain the growth of informal/illegal land transfers in the Global South, it is important to note other factors. Population growth, the trend to commodification and individualisation of customary land, the ineffectiveness of government land administration agencies and an absence of suitable state-led land reform that enables private dealings in customary land are also driving this trend. Regarding the latter point, government efforts in land reform often lag behind what is happening informally as customary landowners and migrants develop their own systems of tenure for land transactions regardless of state policies (Benjaminsen & Lund 2003;

Chimhowu & Woodhouse 2006; Peters 2007; Yaro & Abraham 2009; Yaro 2010). In Fiji, where customary land was being sold to Fijian Indians despite the prohibitions of both law and custom, proposed changes to land laws have 'lagged well behind practice' (Ward 1995, p. 247). In the absence of an effective land administration system in PNG, landowners seeking to capitalise on the demand for land by outsiders are developing their own informal arrangements for land mobilisation (Curry & Koczberski 2009; Koczberski et al. 2009). This is occurring not only in the oil palm regions of PNG but also in rural areas of the Gazelle Peninsula (see Curry et al. 2007b; Martin 2007; Fingleton 2008) and in urban and peri-urban locations (Fingleton 2008; Chand & Yala 2012; Numbasa & Koczberski this issue).

The rapid growth in informal and illegal land transactions presents a challenge for policy makers: how to develop an effective reform program and land administration system to accommodate the wide range of informal and semi-formal arrangements occurring in areas where demand for customary land by outsiders is high. Whilst land reform has long been the subject of policy debate in many developing countries, much of the discussion has been dominated by the view that customary tenure is incapable of providing the necessary security to facilitate investment and productive uses of land (e.g. Gosarevski et al. 2004; IIED 2005; Peters 2007). Thus a key driver of land reform in PNG, and in other parts of the developing world, has been based on the notion that secure individual property rights through titling of land are a fundamental prerequisite for building a modern economy and stimulating economic growth (Gosarevski et al. 2004). This 'individualisation' approach emerged from Africa half a century ago, and it still has wide support, despite the evidence that full individualisation and registration of customary tenures has often failed to deliver the anticipated increase in agricultural investment and productivity (Mackenzie 1993; Shipton 1994; Lastarria-Cornhiel 1997; Lund 2000; Peters 2007; Fingleton 2008).

Recently, however, there has been growing recognition amongst policy makers in PNG, other Pacific Island nations and in many African countries that land reform should aim to support and build on existing customary tenure to meet contemporary needs and demands rather than replace it. This has grown out of the adaptation approach to land tenure reform that was promoted at the 2004 FAO (Food and Agriculture Organization) Land in Africa conference (Fingleton 2008). This 'adaptation' approach to land tenure reform is more likely to be acceptable to customary landowners. As Fingleton (2008) noted in a paper on land reform in the Pacific:

> there is now a general acceptance that adaptation, not replacement, of customary tenures is the way forward. The Food and Agriculture Organization (FAO) of the United Nations endorses the adaptation approach to land tenure reform. Even the World Bank, for long a critic of customary tenures, has given ground, now recognising customary tenures as a viable basis for growth and development. At the Land in Africa Conference, held in London in November 2004, the adaptation approach was given strong support by all the governments and aid agencies that took part. (2008, pp. 10–11)

Thus, contemporary approaches to land reform in the Pacific now seek to develop adaptation models that balance the goal of providing security to investors/ individuals to facilitate economic development on customary land while at the same time protecting existing customary tenure systems and the underlying rights of landowners. This paper presents a template developed recently for a new Land Usage Agreement (LUA) for use among oil palm smallholders. It is argued that the LUA presents a way forward for land reform that builds on customary tenure while strengthening the temporary use rights of migrants to enable them to generate viable and relatively secure livelihoods.

We begin with a discussion of the types of land transactions occurring in the oil palm areas of PNG and explore some of the attendant problems. Then we report on our ongoing research with agricultural extension services, customary landowners and migrant farmers to develop a template for a more formal contractual LUA that seeks to accommodate customary land tenure while providing migrant farmers with secure use rights to land. The new LUA shows a way forward that builds on customary tenure while strengthening the temporary use rights of outsiders to enable them to generate viable and relatively secure livelihoods. The paper concludes with a discussion of the implications of the case study for land reform policy for rural and urban areas undergoing rapid economic and demographic change, and where increasing numbers of people, without customary or formal use rights to land, are accessing land for residence and livelihood activities.

Background to customary land and land reform in PNG

Customary tenure in PNG

Customary land tenure in PNG has evolved to accommodate a diverse array of physical and social environments. Despite the diversity of land tenure regimes, there were two common sets of general principles across most of the country. First, exclusive individual ownership of land was rare as nearly all land was vested in landholding groups grounded in kinship (Ward & Kingdon 1995). Second, principles of customary land tenure were pragmatic, which ensured most people had sufficient access to land to meet their livelihood requirements (food, materials for shelter and cultural needs). Without direct inheritance of individual land rights (because land reverted to the group during fallow periods), the changing demographic fortunes of individual lineages were able to be accommodated (Crocombe & Hide 1971; Curry 1997; Ward 1997).

However, the cultivation of perennial export tree crops such as oil palm, cocoa and coffee induced de facto changes in land tenure (see Epstein 1969; Foster 1995; Curry *et al.* 2007a; Martin 2007). The long-term cultivation of perennial crops has meant that exclusive rights over land planted to export crops are now held by individuals for long periods, thereby reducing the flexibility and pragmatism of land tenure that characterised swidden systems. Yet, despite these changes, land continues to be fundamental to household food production, and for sustaining spiritual beliefs, social and ritual activities, individual and group social identities, and for underpinning social organisation. This explains why in many areas of PNG the notion that land can be alienated permanently is rarely comprehended or internalised by customary landowners.

Land reform in PNG

With over 90 per cent of land in PNG under customary tenure, attempts at land reform to facilitate investment, trade and other entrepreneurial activities have long been perceived by policy makers to be critical to economic development. In the colonial and post-colonial periods various land reform measures were promulgated to encourage the shift from communal tenure to individual leasehold or freehold title. Three major programs to register land and facilitate individual freehold title have been initiated over the years (Kalinoe 2008). These are, first, the *Land (Tenure Conversion) Act* of 1963 to permit the transfer of customary land to individual freehold title. Land tenure conversion was based on the assumption that the individualisation of title would promote individual family farms, thereby creating the conditions for the emergence of commercial agriculture (Morawetz 1967). Land Tenure Conversion (LTC) blocks were promoted in the PNG Highlands to facilitate development of the smallholder coffee industry, and in Popondetta, Oro Province, where village oil palm growers were encouraged to convert their land from customary control to individual freehold title. LTC blocks were considered an intermediary step towards freehold title registration (Morawetz 1967). However, few, if any, of the approximately 450 ha of village oil palm in Popondetta registered as LTC blocks were converted to freehold title.

The second attempt at land reform was the lease, lease-back scheme which commenced in 1979 to facilitate the extension of credit for agricultural developments on customary land (Filer 2011). It was first used in the coffee industry with limited success, and has since been adopted by forestry ventures and more recently by the oil palm industry for plantation estates. Apart from oil palm, the lease, lease-back has not resulted in extensive areas of customary land becoming available for commercial agricultural development. Indeed, recent reports of abuse of the lease, lease-back system have undermined trust in the Lands Department's capacity to effectively manage these leases (Filer 2011).

The most recent attempt at land reform was the Land Mobilisation Program (LMP) which began in 1987 in the East Sepik Province to promote registration and development of customary land. The strategy has been promoted repeatedly since the mid-1980s. The East Sepik LMP was based on two related Acts the *Provincial Lands Act* and the *Customary Land Registration Act* (Larmour 1991) which together allowed for recognition of group 'ownership' and various tenure arrangements on customary land such as joint ventures between Incorporated Land Groups (ILGs) and investors, and leases to individual clan and non-clan members. These reforms acted to 'free up' and facilitate land mobilisation for economic development and other land use requirements in the province (Fingleton 1991; Larmour 1991; Kalinoe 2008). The East New Britain provincial administration experimented with a similar LMP, and in the 1990s a National Land Mobilisation Program to register customary land was witness to public condemnation and rioting (1995 and 2001), resulting in land registration in PNG largely coming to a halt.

Each of the three land reform programs discussed above has had limited success (Larmour 1991; Fingleton 2004; National Research Institute 2007). Some have attributed their failure to the weak institutional frameworks and a lack of support at the regional and state levels to develop and enforce a formal property rights regime that provides security for access to resources and capital investment (e.g. Jones & McGavin 2000). However, it is not only institutional weakness at the macro level that explains the limited success of these measures but also their incompatibility at the village level with indigenous concepts of the moral basis of land rights and the emphasis on relational identities for land access (discussed further below) (Curry & Koczberski 2009).

Land reform is once again on the policy agenda in PNG, with a National Land Summit in 2005 and a National Land Development Taskforce (NLDT) being instituted in the same year. One of the central goals of the NLDT was to identify ways to facilitate access to customary land for economic development which provided security for individuals and investors while ensuring customary land remained under the ownership of customary landowners (National Research Institute 2007). The specific type of land tenure reform required to achieve these twin objectives is still under discussion, but it is clear that contemporary approaches to land reform in PNG are becoming more adaptive. This adaptation approach was reflected in the overall goal of the 2005 reform program of the Customary Land Development Committee of the NLDT which proposed that 'customary land will remain in the possession (ownership) of the landowning group, yet can be comfortably leased and utilized freely, in the modern business environment' (National Research Institute 2007, p. xvi).

Field sites and methods

Commercial production of oil palm began along the north coast of WNB Province in 1968 and in Oro Province in 1976 with state acquisition and conversion of customary land to state agricultural leases for the establishment of government land settlement schemes (LSSs) and plantation estates (see Figure 1). Families, largely from mainland PNG, were voluntarily resettled on the schemes at Hoskins and Bialla, WNB and in Popondetta, Oro Province, and allocated individual 99-year state agricultural leases over landholdings of 6.0–6.5 ha. Following the establishment of the oil palm LSSs, neighbouring customary landowners living on their village lands began planting oil palm. Later, in WNB, some landowners began 'selling' customary land to outsiders to plant oil palm, and over 3500 ha of customary land in the Hoskins and Bialla areas have been 'sold' to outsiders, many of whom are second-generation LSS settlers or employees of companies or government seeking to retire in the province and/or secure a future for their children (Koczberski et al. 2009). The oil palm LSSs are one of the few cases in PNG of significant long-term rural-to-rural migration. The success of the oil palm industry (the leading export cash crop by value) is dependent on workable land tenure regimes that are able to accommodate outsiders (Koczberski et al. 2001).

Data were collected from 2006 to 2010 from numerous meetings, informal interviews and workshops with members of landowning groups involved in the 'sale' of customary land, migrant oil palm smallholders cultivating oil palm on customary land, and agricultural extension officers of OPIC. Initially, the research documented the types of land transactions occurring and how tenure and access rights were acquired by migrants and maintained through time. This involved examining landowner and migrant understandings of land transactions and the underlying causes of the rising number of land disputes. As it became apparent that landowners and migrants had different interpretations of their land dealings and both sought more formal processes for land transactions, including written documentation of the rights and obligations of both parties, work began with landowning groups, migrants and OPIC to develop an LUA template. This was an iterative process over several years in which the initial concerns of both migrants and customary landowners about land transactions were documented and brought back to both groups for further discussion and refinement.

Changing land tenure regimes in the oil palm growing areas of West New Britain

Land tenure regimes have changed considerably in the oil palm belt of WNB in response to the high demand for land by migrants for oil palm and non-oil palm livelihoods (see Table 1). Although Table 1 uses the dichotomy of customary and alienated land to describe the two categories of legal tenure in PNG, it is evident that the different tenure arrangements within each category challenge this simple division. For example, in an environment where national institutional governance is weak, state land is being appropriated by migrants and 'reclaimed' by those who claim to be customary landowners. Also, the principles of customary land tenure are being manipulated by clan members to allow the temporary/partial alienation of land through land 'sales' leading to the emergence of new property and social relations. Thus, a wide range of overlapping tenure regimes and arrangements have emerged as ideas of individual and communal ownership and property relations have been reworked on both alienated and customary land (see Table 1).

The 'sale' of land to outsiders and rising discontent

When migrants 'purchased' customary land for oil palm cultivation, a 'price' for the land was agreed upon, although transactions usually tended to be informal with ambiguous verbal or written agreements made between the outsider and customary landowners.[1] Land surveys were rarely undertaken and written agreements often did not specify the agreed 'sale' price of the land, the amount and timing of payment instalments, and the specific rights and obligations of the migrant or the landowners. Seldom was there any written evidence that the land transaction had the approval of the clan. Disagreements could arise within the landowning group when some members had not consented to the land transaction or did not receive a share of the cash from the transaction. Such disputes sometimes led to the eviction of migrants who found little recourse through the courts because of the lack of written evidence regarding the land transaction.

Disputes between migrants and customary landowners often arose because of the conflicting interpretations of land transactions (Koczberski *et al.* 2009). Many migrants acquiring land interpreted the land transfer as a commodity transaction that conferred on them individual and permanent ownership (like freehold title). From this perspective the land was to be held in perpetuity by the migrant which therefore gave his sons inheritance rights or allowed him to on-sell the land to a third party. Those migrants who discursively constructed the land transfer as a commodity transaction also believed that they were free to pursue other livelihood activities on the land, such as operating small business enterprises, and that there were no obligations on them to share the wealth generated from the land with the landowners or to maintain social relationships with them.

Landowners, in contrast, drew on customary principles that viewed land as an inalienable resource held by the kinship group, a perspective that was held by most members of the landowning group (Curry & Koczberski 2009). As customary

TABLE 1. The types of formal and informal tenure systems operating in the oil palm belt of West New Britain (WNB)

Tenure system	Description
Alienated leasehold land	
State agricultural leasehold land	Oil palm smallholder land settlement schemes and plantation estates on 99-year state agricultural leases. Estimated area of state agricultural leases planted to oil palm in WNB is 43 280 ha
Formal and informal reclamation of state agricultural leasehold land	Informal reclamation of state land by landowning groups and individual members of the landowning group. Some reclaimed state land is 'rented' or has been 'sold' illegally to migrants. Migrants also known to have appropriated state agricultural leasehold land
State land in urban areas	Informal urban and peri-urban settlements illegally occupied by migrants
Lease, lease-back system (since 1996)	Registered sub-leases of 20–40 years between customary landowning groups and the milling company for the development of oil palm plantation estates on customary land. The lease suspends customary land use. Over 36 000 ha are under lease, lease-back arrangements in WNB
Customary land	
Village Oil Palm (VOP) holdings	Individual clan members' oil palm blocks on customary land and 'registered' with the Oil Palm Industry Corporation. Some VOP blocks are under 'Clan Land Usage Agreements' and some are occupied by village non-clan members by arrangement with customary landowners
Community-managed oil palm plantations (Bialla project area)	Amalgamations of clan land. Approximately 6500 ha under Community Oil Palm plantations
Informal and semi-informal transfers of customary land to non-clan members	Transfers are largely 'sales' of land to 'outsiders' to plant oil palm. Transfers of land are sometimes 'gifted' to outsiders to allow short-term or long-term access to small parcels of customary land
'Renting' of customary land to migrants	Informal rental agreements give migrants access to land for food gardening and house sites in return for cash and/or labour

landowners they believed they had the power to determine what livelihood activities migrants could pursue on the land. They argued that outsiders had acquired only use rights for oil palm and not rights to pursue other income-earning activities. Also, for landowners, land rights granted to outsiders were not permanent and exclusive. Instead, a less exclusive set of rights pertained that were contingent on migrants' continued participation in indigenous exchange and fulfilling other social and economic obligations—expectations similar to more traditional concepts of land tenure where access rights granted to outsiders for subsistence gardens depend on the maintenance of social and economic relationships with landowners.

These 'guests' on customary land were expected to share a portion of their income from oil palm or other income activities on the land and contribute to community and village events. Ideally, income sharing with hosts should reflect the benefits that migrants were deriving from the land. When oil palm prices rose significantly, some landowners felt they were being cheated because they 'sold' the land when oil palm prices were lower. This led some to believe that migrants were receiving benefits unfairly from the higher prices which could induce landowners to demand additional payments from migrants. When landowners believed outsiders had failed to maintain exchange relationships adequately, the moral basis of a migrant's land claim weakened, and there was a corresponding strengthening of the customary landowners' own moral claims to the land. For landowners, the transactions were very much interpreted in terms of indigenous principles of land tenure that stressed the inalienability of land and the relational basis of land rights.

Elements of a template for a new clan Land Usage Agreement

The differing interpretations of land transactions by landowners and migrants, together with the associated rising incidence of land disputes, indicated a need to review current practices regarding the transfer of customary land to outsiders. Moreover, as mentioned above, landowners and migrants both sought more formal land transaction processes and OPIC increasingly acknowledged that the existing informal arrangements and LUAs used for customary land transactions with outsiders for the planting of oil palm did not provide adequate tenure security for outsiders, and nor did they ensure that all members of the landowning group consented to, or benefited from, these land transactions. Thus, following wide consultations with landowners and migrants, the important elements of an LUA emerged which formed the basis of a template that migrants and landowners could modify to meet their own particular needs and circumstances.

There were two broad aims in developing a template for a LUA. The first aim was to create a more transparent process in accordance with customary law, so that all individuals with a stake in the land would be brought into the process of negotiation. Second, given that landowners and migrants interpreted land transactions differently, there was a need in these negotiations to reconcile their disparate perspectives with the objective of giving greater tenure security to outsiders while recognising the underlying tenure rights of the landowning group. We did not seek to be proscriptive in developing a template but rather sought to bring into discussions matters that potentially could lead to disputes if they were not considered during initial negotiations.

Transparent transactions

The main components of the LUA template were concerned with making the land transaction more transparent and developing clearer definitions of the rights and obligations of both parties. The main point of agreement amongst landowners was that the template should state explicitly that the outsider acquiring the land was not purchasing the land outright as in freehold title, but rather obtaining usufruct rights to the land for defined purposes (oil palm production, food gardening for household consumption and family residence) and for a specified period of time.

Their objective was to ensure that the underlying customary rights of the landowning group were preserved.

Transparent land transactions were also important for migrants. Many migrants attributed the discontent of the younger generation of landowners towards migrants to the fact they were not party to land transactions and some had no knowledge that land had been transferred until it was being cleared by migrants. To enhance transparency, a three-party signatory process was included in the LUA template which comprised the landowning clan members (at least four senior clan members, including women leaders), individuals from the clan who, under customary law, had use rights to the land, and the outsider acquiring the land. These signatures were to be witnessed by the OPIC Officer, the local government Land Mediator, the Ward Councillor, a local church leader and a clan leader from a neighbouring landowning group. The agreement was to be signed publicly, with good representation from the landowning group present, including those individuals or families authorised and identified by the community to deal in land matters. Prior to signing the LUA, confirmation would have to be made that a majority of landowners with an interest in the land gave free prior and informed consent and supported the proposed land transaction. If the majority opposed the transaction, the agreement was not to be signed and the transaction cancelled. In reality, the consensus form of local governance that is common in rural PNG meant that eventually everyone would reach a unanimous decision as to whether or not the transaction should proceed. To complete the transaction, a block inspection report was to be undertaken which identified the land parcel and confirmed its suitability for oil palm. Rather than undertaking an expensive land survey of the block, the cost of which was beyond the means of most smallholders, land boundaries were to be surveyed by GPS and clearly demarcated using traditional means such as planting along plot boundaries with *Cordyline* (*tanget*) species, coconut or betel nut palms.

Specific rights and obligations

Other rights and obligations of the clan leaders and clan members disposing of the land, and the outsider, that emerged during discussions, and which were incorporated into the template, included clarification that the landowners relinquished any use or management rights to the land for the duration of the agreement and stipulated the specific rights of, and restrictions on, migrants. For example, the rights of the migrant to plant other cash crops and food crops, to establish small businesses, houses and other assets and to bury their dead on the block were often major sources of conflict between landowners and migrants. For instance, burials of migrants' family members on the block were opposed by all landowners because such practices were a way to assert permanent land rights. As an alternative, the LUA specified that burials could take place on prescribed village cemeteries free of charge. These rights and obligations were specified in the template.

Similarly, most landowners were opposed to members of the migrant's extended family residing on the block. Landowners pointed to blocks that were like small villages because so many of the migrant's relatives were living there. The default option in the LUA was that only the migrant and his/her immediate family could reside permanently on the block. Interestingly, several migrants were not opposed to the restrictions on who could settle with them on the block.

Migrants saw it as a way to prevent relatives from residing with them indefinitely, and it gave them an acceptable excuse to avoid what could become onerous financial obligations.

Regarding the land itself, all landowners were opposed to migrants on-selling or sub-leasing the land to a third party or allowing migrants' relatives to manage oil palm holdings. Landowners argued they had the right to repossess the land and return it to common property because in such cases migrants were seen as forfeiting their rights to the land because they were terminating the relationship with their host lineage upon which land access depended. From the perspective of the landowners, because the new occupants of blocks did not have relationships with landowners, they did not have a moral claim to occupy the land. Similarly, some landowners were of the view that oil palm blocks should pass back into customary ownership if migrants were to die before the LUA expired, the rationale being that transactions were only with migrants, not their families. However, this view was uncommon, and most landowners agreed to include a clause in the template that the immediate family, typically the spouse, should inherit the block for the remaining period of the LUA.

Payments and compensation

Another topic discussed extensively was payment for use rights to the land. Customary land 'sells' for between K500 and K3000 per hectare (AUS$1.00 = K2.50, Westpac exchange rate 20 September 2011), and payment is made in instalments over several years, although no formal agreements are usually in place for this. Landowners requested, as part of the new LUA, that alongside the initial payment for the land, migrants should make regular payments to the landowning group to retain long-term access to the land. This ongoing payment was to be based on the value of production from the oil palm holding, similar to a royalty, and applied for the duration of the agreement. Thus, as production from the block increased through time and/or oil palm prices rose, landowners would receive a share of the increase in the value of production. Similarly, if oil palm prices fell, landowners too would share the pain of migrants as their returns would also fall. A production fee of 5–10 per cent of the value of production was considered an appropriate rate. In line with the emphasis on transparency, both parties agreed that all payments and payment arrangements should be transparent and documented, preferably by automatic deductions from growers' payments.

A final area for inclusion in the LUA template, and which was raised by migrants, related to compensation for non-removable assets such as coconut/betel nut palms and water tanks if the LUA were not renewed on expiry. One suggestion for handling compensation for non-removable assets involved the suspension of the oil palm production fee for the last 5 years of the agreement. Deductions would continue but be held in trust until it was decided whether or not the LUA would be renewed. If it were not renewed the lump sum would be paid to the migrant as compensation for vacating the block; if it were renewed, the lump sum would be paid to the landowners and considered to be *hamamas pei*[2] (see below), that is, a gift to renew the relationships for a another planting round of oil palm.

Indigenising land transactions—nurturing social relationships

To further enhance the transparency of land transactions and legitimise them in terms of customary practices, landowners and some extension officers recommended that land transfers should be indigenised rather than simply mimic Western ways of dealing in land which had less legitimacy in the eyes of most villagers because they were not relational based. In Oro Province, for instance, many of the early oil palm blocks in the 1970s and 1980s were established on land gifted to fellow villagers. Often these land gifts were reciprocated with large feasts hosted by the person receiving the land at which cash, pig meat and large quantities of other foods were presented to customary landowners. These large celebratory feasts were attended by most members of the host landowning group as well as by clan leaders from neighbouring groups who witnessed the land transaction and the gifts received by the customary landowning group (Curry & Koczberski 2009; similar processes for legitimising land rights have been noted in Africa, e.g. Mathieu *et al.* 2002; Chauveau 2006; Chimhowu & Woodhouse 2006). Large-scale communal feasting and gift giving thus legitimised and gave public recognition to the moral rights of the outsider to produce oil palm on the land of his 'host'.

By constructing these land dealings as indigenous exchange transactions they are excised from the realm of commodity exchange and given a moral underpinning amongst landowners that serves to strengthen the tenure security of migrants acquiring the land. Embedding the land transaction in large-scale communal feasting and gift exchange established three critical points for establishing a moral basis to land rights: (1) that the transaction and its meaning is transparent and apparent to all; (2) by partaking in the feast, participants are acknowledging the transfer of land rights to the outsider and endorsing the latter's access rights to the land; and (3) that migrants' land rights are relational based, a relation created by communal feasting and gift exchange. It therefore becomes difficult for landowners to renege on these agreements—there are simply too many witnesses to the moral basis of the land rights of the outsider. Communal feasting and exchange also ensures that migrants acquiring land are drawn into these exchange relationships and therefore know and understand their future obligations for maintaining these relationships and, by extension, their ongoing access rights to land.

Similarly, whilst the notion of a production fee mentioned above received universal support amongst landowners, some migrants were opposed to this idea. These migrants held the view that their acquisition of land was a commodity transaction which entailed no further obligations to and relationships with the customary landowners beyond initial payments. However, a significant proportion of migrants realised that the absence of an ongoing relationship with landowners was often the principal reason for the disputes and strained relationships with their hosts. The proposed production fee draws on indigenous value registers where the bounty of the block is shared with the host lineages in ways that enhance and strengthen the social relationships between them (Curry & Koczberski 2012). The production fee is similar to payments in indigenous exchange for granting outsiders access to land for food gardening, and again places the relationship within the realm of indigenous exchange thereby enhancing the moral legitimacy of outsiders' land rights.

Conclusions and implications for policy reform

Like other landowners in developing countries, particularly in Africa (Peters 2007; Yaro 2010), landowners in PNG are keen to realise the economic opportunities on their land, and have demonstrated their desire to do so by entering a diverse array of tenure arrangements with outsiders. However, as discussed in the introduction, land reform programs based on Western models of land tenure have had limited success in PNG, and increasingly policy makers are acknowledging that land reform should aim to support and build on existing customary tenure rather than replace it. The LUA template described in this paper accords with this shift to an adaptation approach to land reform. One of the central goals of the new LUA was to develop a template that maintained customary ownership at the group level but strengthened the use rights of individuals for the full cultivation cycle of oil palm. This endorses Fingleton's recommendation that a 'two-tier registration system, with group titles as the "head title" (i.e. ownership), and then subsidiary titles (such as leases) granted by groups to the users of the land' (2005, p. 35) is a feasible reform.

By designing an LUA template in collaboration with landowners and outsiders, the case study has three implications for policy reform. First, the ways in which customary tenure has been adapted, manipulated and modified in response to migrants' demand for land demonstrates the capacity of customary tenure to accommodate new challenges and generate 'new forms out of old' for the modern context. It shows that solutions are to be found in what is already happening on the ground, rather than in externally imposed models that do not fit well with local cultural mores about the inalienability of customary land and its enduring social and cultural significance for customary landowning groups.

Second, and related to the first point, while landowners are eager to realise the income potential of their land, they are only keen to do so if their customary tenure rights are preserved. The LUA template provides use rights for outsiders for fixed time periods with clear recognition of the underlying and inalienable land rights of customary landowners. Thus, as recently proposed by the NLDT, adaptation of existing land tenure systems rather than wholesale change is a way forward. Indeed, the informal land transactions already taking place in WNB illustrate to a large extent that it is not necessary to convert customary land to individual freehold title to improve economic performance of the smallholder sector, as argued by some (e.g. Curtin & Lea 2006; Gosarevski et al. 2004). Despite their insecure and temporary tenure, migrants cultivating oil palm on customary land have very high productivity. Their production levels are much higher than those of landowners and are comparable with smallholders residing on state agricultural leasehold blocks registered to individual farmers (unpublished data).

Third, land use agreements must be based on a clear understanding of the local socio-cultural and environmental contexts. One of the key elements of the LUA template is that it draws on indigenous principles and relational concepts of property rights. Thus, the search for potential models to mobilise land are not necessarily to be found in externally imposed models that treat land as an alienable commodity. Potential solutions and models can be derived by drawing on what is already happening at the local level and the adaptations and modifications that are already taking place on customary land outside government structures as landowners and migrants themselves develop solutions to meet new circumstances. These innovations draw on customary principles for their validity and moral basis, and the goal of policy should be to strengthen these aspects rather than undermine them.

Acknowledgements

The research was funded by the Australian Centre for International Agricultural Research. The research was also informed by fieldwork conducted as part of an ARC Discovery Project on migrant land issues in WNB. The authors are grateful for the insightful feedback of the referees.

NOTES

[1] Section 81 of the *Land Act* prohibits the sale of customary land except to citizens of PNG in accordance with customary law.
[2] *Hamamas pei* is a form of gift exchange that refers to 'payments' to elicit positive feelings towards the giver by the receiver of the payment.

REFERENCES

AMANOR, K. & DIDERUTUAH, M. (2001) *Share contracts in the oil palm and citrus belt of Ghana*, International Institute for Environment and Development, London.

BENJAMINSEN, T.A. & LUND, C. (2003) *Securing land rights in Africa*, Frank Cass, London & Portland.

CHAND, S. & YALA, C. (2012) 'Institutions for improving access to land for settler-housing: evidence from Papua New Guinea', *Land Policy* 29(1), pp. 143–53.

CHAUVEAU, J.P. (2006) 'How does an institution evolve? Land, politics, intergenerational relations and the institution of the *Tutorat* amongst autochthones and immigrants (Gban region, Côte d'Ivoire)', in Kuba, R. & Lentz, C. (eds) *Land and the politics of belonging in West Africa*, Koninklijke Brill, Leiden, pp. 213–41.

CHIMHOWU, A. & WOODHOUSE, P. (2006) 'Customary vs private property rights? Dynamics and trajectories of vernacular land markets in sub-Saharan Africa', *Journal of Agrarian Change* 6(3), pp. 346–71.

CROCOMBE, R. & HIDE, R. (1971) 'New Guinea: unity in diversity', in Crocombe, R. (ed.) *Land tenure in the Pacific*, Oxford University Press, Melbourne, pp. 292–333.

CURRY, G.N. (1997) 'Warfare, social organisation and resource access amongst the Wosera Abelam of Papua New Guinea', *Oceania* 67(3), pp. 194–217.

CURRY, G.N. & KOCZBERSKI, G. (2009) 'Finding common ground: relational concepts of land tenure and economy in the oil palm frontier of Papua New Guinea', *Geographical Journal* 175(2), pp. 98–111.

CURRY, G.N. & KOCZBERSKI, G. (2012 in press) 'Relational economies, social embeddedness and valuing labour in agrarian change: an example from the developing world', *Geographical Research* 50 doi: 10.1111/j.1745-5871.2011.00733.x

CURRY, G.N., KOCZBERSKI, G., OMURU, E., DUIGU, J., YALA, C. & IMBUN, B. (2007a) *Social assessment of the Smallholder Agriculture Development Project*, report prepared for the World Bank.

CURRY, G.N., KOCZBERSKI, G., OMURU, E. & NAILINA, R.S. (2007b) *Farming or foraging? Household labour and livelihood strategies amongst smallholder cocoa growers in Papua New Guinea*, Black Swan Press, Perth.

CURTIN, T. & LEA, D. (2006) 'Land titling and socioeconomic development in the South Pacific', *Pacific Economic Bulletin* 21(1), pp. 153–80.

DALEY, E. (2005) 'Land and social change in a Tanzanian village 2: Kinyanambo in the 1990s', *Journal of Agrarian Change* 5(4), pp. 526–72.

ELMHIRST, R. (2001) 'Resource struggles and the politics of place in North Lampung, Indonesia', *Singapore Journal of Tropical Geography* 22(3), pp. 284–306.

EPSTEIN, A.L. (1969) *Matupit: land, politics and change among the Tolai of New Britain*, Australian National University Press, Canberra.

FILER, C. (2011) *The political construction of a land grab in Papua New Guinea*, Resources, Environment and Development (READ) Pacific Discussion Paper 1, Crawford School, Australian National University, Canberra.

FINGLETON, J. (1991) 'The East Sepik land legislation', in Larmour, P. (ed.) *Customary land tenure: registration and decentralization in Papua New Guinea*, Monograph No. 29, Papua New Guinea Institute of Applied Social and Economic Research, Boroko, pp. 147–62.

FINGLETON, J. (2004) 'Is Papua New Guinea viable without customary groups?', *Pacific Economic Bulletin* 19(2), pp. 96–103.

FINGLETON, J. (2005) 'Conclusion', In: Fingleton, J. (ed.) *Privatising land in the Pacific: a defence of customary tenures*, Discussion Paper Number 80, Australia Institute, Canberra, pp. 34–7.

FINGLETON, J. (2008) *Pacific land tenures: new ideas for reform*, FAO Legal Papers Online #73, available from: www.fao.org/legal/prs-ol (accessed 1 March 2012).

FOSTER, R.J. (1995) *Social reproduction and history in Melanesia: mortuary ritual, gift exchange, and custom in the Tanga islands*, Cambridge University Press, Cambridge.

GOSAREVSKI, S., HUGHES, H. & WINDYBANK, S. (2004) 'Is Papua New Guinea viable?', *Pacific Economic Bulletin* 19(1), pp. 134–48.

HAGBERG, S. (2006) 'Money, ritual and the politics of belonging in land transactions in Western Burkina Faso', in Kuba, R. & Lentz, C. (eds) *Land and the politics of belonging in West Africa*, Koninklijke Brill, Leiden, pp. 99–118.

IIED (INTERNATIONAL INSTITUTE FOR ENVIRONMENT AND DEVELOPMENT) (2005) 'Land in Africa market asset or secure livelihood?', Issue Paper No. 136, Royal African Society, London.

JONES, L.T. & MCGAVIN, P.A. (2000) *Creating economic incentives for land mobilisation in Papua New Guinea: a case study analysis of the formation and maintenance of institutions that assist mobilisation of land for agricultural uses*, Institute of National Affairs, Port Moresby.

KALINOE, L. (2008) *Review of incorporated land groups and design of a system of voluntary customary land registration*, Report 5, Constitutional and Law Reform Commission, NCD, Boroko.

KOCZBERSKI, G., CURRY, G.N. & GIBSON, K. (2001) *Improving productivity of the smallholder oil palm sector in Papua New Guinea*, Department of Human Geography, Research School of Pacific and Asian Studies, Australian National University, Canberra, available from: http://espace.lis.curtin.edu.au/archive/00000235/ (accessed 1 March 2012.).

KOCZBERSKI, G., CURRY, G.N. & IMBUN, B. (2009) 'Property rights for social inclusion: migrant strategies for securing land and livelihoods in Papua New Guinea', *Asia Pacific Viewpoint* 50(1), pp. 29–42.

LARMOUR, P. (1991) 'Registration of customary land: 1952–1987', in Larmour, P. (ed.) *Customary land tenure: registration and decentralization in Papua New Guinea*, Monograph No. 29, Papua New Guinea Institute of Applied Social and Economic Research, Boroko, pp. 51–72.

LASTARRIA-CORNHIEL, S. (1997) 'Impact of privatization on gender and property rights in Africa', *World Development* 25(8), pp. 1317–33.

LI, T.M. (2002) 'Local histories, global markets: cocoa and class in upland Sulawesi', *Development and Change* 33(3), pp. 415–37.

LUND, C. (2000) 'African land tenure: questioning basic assumptions', Drylands Programme Issue Paper No. 100, IIED, London.

MACKENZIE, F. (1993) '"A piece of land never shrinks": reconceptualizing land tenure in a small-holding district', in Basset, T.J. & Crummey, D.E. (eds) *Land in African agrarian systems*, University of Wisconsin Press, Madison, pp. 194–221.

MARTIN, K. (2007) 'Land, customary and non-customary, in East New Britain', in Weiner, J.F. & Glaskin, K. (eds) *Customary land tenure and registration in Australia and Papua New Guinea: anthropological perspectives*, Asia-Pacific Environment Monograph 3, ANU E Press, Australian National University, Canberra, pp. 39–56.

MATHIEU, P., ZONGO, M. & PARÉ, L. (2002) 'Monetary land transactions in Western Burkina Faso: commoditisation, papers and ambiguities', *European Journal of Development Research* 14(2), pp. 109–28.

MORAWETZ, D. (1967) *Land tenure conversion in the Northern District of Papua New Guinea*, Research Bulletin 17, New Guinea Research Unit and Australian National University, Port Moresby and Canberra.

NATIONAL RESEARCH INSTITUTE (2007) *The National Land Development National Taskforce report: land administration, land dispute settlement and customary land development*, Monograph 39, a report prepared by the NLDT Committees on Land Administration, Land Dispute Settlement, and Customary Land Development, Boroko.

NUMBASA, G. & KOCZBERSKI, G. (this issue) 'Migration, informal urban settlements and non-market land transactions: a case study of Wewak, East Sepik Province, Papua New Guinea', *Australian Geographer* this issue.

PETERS, P.E. (2007) 'Challenges in land tenure and land reform in Africa: an anthropological perspective', Working Paper No. 141, Centre for International Development, Harvard University, Cambridge, MA.

POTTER, L. & BADCOCK, S. (2004) 'Tree crop smallholders, capitalism, and *Adat*: studies in Riau Province, Indonesia', *Asia Pacific Viewpoint* 45(3), pp. 341–56.

SHIPTON, P. (1994) 'Land and culture in tropical Africa: soils, symbols, and the metaphysics of the mundane', *Annual Review of Anthropology* 23, pp. 347–77.

SJAASTAD, E. (2003) 'Trends in the emergence of agricultural land markets in sub-Saharan Africa', *Forum for Development Studies* 30(1), pp. 5–28.

WARD, R.G. (1995) 'Land, law and custom: diverging realities in Fiji', in Ward, R.G. & Kingdon, E. (eds) *Land, custom and practice in the South Pacific*, Cambridge Asia-Pacific Studies, Cambridge University Press, Cambridge, pp. 198–249.

WARD, R.G. (1997) 'Changing forms of communal tenure', in Larmour, P. (ed.) *The governance of common property in the Pacific region*, National Centre for Development Studies, Research School of Pacific and Asian Studies, and Resource Management in Asia-Pacific, Australia National University, Canberra, pp. 19–32.

WARD, R.G. & KINGDON, E. (eds) (1995) *Land, custom and practice in the South Pacific*, Cambridge Asia-Pacific Studies, Cambridge University Press, Cambridge.

YARO, J.A. (2010) 'Customary tenure systems under siege: contemporary access to land in Northern Ghana', *GeoJournal* 75(2), pp. 199–214.

YARO, J.A. & ABRAHAM, Z. (2009) 'Customary land tenure, investment and livelihoods adaptation in Northern Ghana', *Ogua Journal of Social Science* 4(3), pp. 59–82.

Access to Land and Livelihoods in Post-conflict Timor-Leste

PYONE MYAT THU, *College of Asia and Pacific, Australian National University, Australia*

ABSTRACT   *East Timor gained formal independence in 2002. Its extended history of internal displacement through colonial territorialisation strategies and conflict has produced an array of contesting land claims, informal land use and occupation, and socio-political conflict. Correspondingly, the Timor-Leste (or East Timor) government has looked to formal land registration and titling to resolve historical and contemporary tensions over land. The proposed national land laws concentrate on land ownership, which overlooks the social relations at work to shape local land access and livelihoods. A case study of a rural village, Mulia, forcibly resettled during the Indonesian occupation, demonstrates how settlers negotiate access to customary land for livelihoods despite ongoing land conflicts with the customary landowners. Settlers have continually adapted to broader economic and political constraints to create diverse and multi-local livelihoods, and established a moral economy between themselves, landowners and the local spirit realm. This paper argues that formal land titles are unlikely to resolve the ambiguities and complexities of diverse forms of access to and ownership of land.*

Introduction

Post-conflict land issues in Timor-Leste are highly complex. The urgency to resolve land and property disputes in post-conflict Timor-Leste has been prompted by ongoing outbreaks of communal violence after national independence (Gunter 2007; Harrington 2007; Scambary 2009). The Timor-Leste government has therefore sought land titling to address these matters. Formalisation of land rights is advocated primarily by neoliberal policy makers who take an economic view of land as 'dead capital' (see, for example, De Soto 2000). The mobilisation of 'dead capital' through land titles, so demarcating clear and legally enforceable rights, is posited to clarify competing interests in land, enable landholders to exercise greater control of land, and free up land for productive investment and collateral loan (De Soto 2000; Hughes 2004). In post-conflict settings, such as in the case of Timor-Leste, the need for clear and enforceable land rights is further justified on the grounds that they will bring peace, reconciliation, post-war reconstruction, foreign investment, and sustainable livelihoods (Du Plessis 2003; Fitzpatrick 2002; Ita Nia Rai 2005; Unruh 2008).

A growing body of evidence from South Asia, Southeast Asia, Africa, and Latin America has shown, on the contrary, that formal land titles do not always guarantee security of tenure, protection from land grabs, poverty alleviation, and gender equity (Agarwal 2003; Manji 2006; Lund 2000; Hall *et al.* 2011). Taking a broader view, Hall *et al.* (2011, p. 27) contend that securing land rights for one group of people inevitably entails the exclusion of others. In post-colonial and post-conflict situations, access to land can be ambiguous given that there are usually coexisting and contesting politico-legal and socio-political institutions that can determine land rights, creating a plural legal context (Meitzner Yoder 2003; Unruh 2008; Sikor & Lund 2009). Further, the post-conflict/post-independence context is usually marked by rapid political, economic, and social changes, implying a kaleidoscope of informal land and property transactions more fluid than in the colonial past. Together, these circumstances question the central role reserved for legal solutions, and stress the need to examine the specific historical, political, institutional, and socio-cultural contexts to better understand how conflict might be managed, land access negotiated, and benefits distributed—in other words, to examine the underlying power relations at work.

Broadly, there are four competing categories of land claims in Timor-Leste: customary tenure, Portuguese titles, Indonesian titles, and Temporary Use Agreements (TUAs) distributed during the transitional period to independence between 1999 and 2002 (Fitzpatrick 2002). Moreover, the full typology of land and property transactions and arrangements is even more complicated, as these claims intersect and overlap with one another, and give rise to different outcomes. In the post-independence era, residues of conflict and colonial territoriality persist in forms of inter-generational violence and social tension manifested in land and property disputes. Conversely, land and property conflicts can also evolve into socio-political differences (Meitzner Yoder 2003; Gunter 2007; Harrington 2007; Scambary 2009). Within this wider context, the Timor-Leste government seeks to advance formal land titling to resolve colonial-inherited land disputes and social conflict.

The national focus on land ownership places emphasis on ownership over access, contestation over cooperation, and exclusion over inclusion. Land titles give undue attention to land, and ignore gender inequalities, remuneration of labour, appropriation of produce and the control of benefits reaped from land (Agarwal 2003; Li 1996; Meinzen-Dick & Mwangi 2009). Inevitably, the range of actors that seek to reap benefits from the land in question is overlooked, and the power relations that shape the ability to derive those benefits remain obscure (Ribot & Peluso 2003). To whom and to what should land policy pay attention? Whose voices are potentially marginalised in enforcing formal titles to land? In seeking to answer these questions, I examine the actual practices of land access and land control under customary land tenure. A focus on *access* rather than *ownership* opens up a space to investigate the situated fields of power that form under local historical contingencies and makes visible the diverse and informal nature of land tenure practices ensuing from internal displacement in rural Timor-Leste.

Taking land access as my focus, I draw on Ribot and Peluso's (2003, p. 153) definition of access as 'the ability to benefit from things'. Access is distinguishable from property; the former is not necessarily limited to socially sanctioned or legalised rights, whilst the latter requires legitimisation from some form of authority—politico-legal or socio-political institutions (Sikor & Lund 2009). Property can therefore be one dimension of access (Ribot & Peluso 2003). Mechanisms of access may be gained through capital, labour, technology, and knowledge, while the conditions that enable access constantly change with the political and economic conditions (Ribot & Peluso 2003, p. 156). I consider in this paper how a group of displaced people have gained access to land in an Indonesian-established resettlement village under customary land tenure. My contention is that without a proper understanding of the social relations that eventuate in access, statutory land rights can perpetuate systemic inequality in power and resource control.

By focusing on how access to land and customary land tenure has changed in rural Timor-Leste due to a legacy of foreign occupation and conflict, this paper also contributes to recent scholarship on the changing nature of land access and land control (Hall *et al.* 2011; Peluso & Lund 2011). Conditions of access and control are changing and becoming more complex through the interactions of new actors and subjects, which have consequently shifted the situated fields of power. Regional comparisons can additionally be made with Southeast Asia, Melanesia and sub-Saharan Africa on the adaptability, responsiveness, and negotiability of customary tenure under the pressures of conflict, commercialisation of agriculture, natural resource extraction and globalisation (Lund 2000; Bainton 2009; Curry & Koczberski 2009; Koczberski *et al.* 2009; Cramb & Sujang 2011). These studies show how customary systems are constantly revising the meanings of property relations, simultaneously redefining social relations and images of 'community', to respond to broader political, economic and ecological conditions.

The first section foregrounds the historical context of displacement and dispossession to chart the broader structural changes that reshaped local access to land and continue to reshape them. In particular, Timor-Leste state's current national development strategy appears to be replicating the colonial processes and practices that resulted in displacement and conflict. The second section focuses on a particular rural locality where a group of displaced people were forced to resettle during the Indonesian occupation to examine how they have negotiated access to land from the customary landowners despite ongoing inter-community conflicts. Here I focus on differentiating land access from land ownership to highlight the social relationships between settlers and landowners, and the power dynamics that shape access. The local conditions of access to land in Mulia must be understood within the broader context of the country's legacy of displacement and disposses-sion. The third section examines the potential implications of the proposed Transitional Land Law for rural land, and the livelihoods that depend on customary land access. I argue that land titles alone will not resolve land conflicts, and that mediation and reparation must be taken into consideration.

Colonial displacement and dispossession

Recent scholarship on post-conflict Timor-Leste has focused on the impacts of internal displacement, namely human rights violations, land disputes and social conflict, rather than seeking to understand the phenomenon itself (e.g. Fitzpatrick 2002; Unruh 2008). Most East Timorese have experienced forced displacement during their lifetimes. As a consequence of an extended history of foreign incursions under Portuguese colonialism (sixteenth century to 1974), the brief Japanese occupation (1942–44), and the Indonesian occupation (1975–99), the overriding narrative of internal displacement in Timor-Leste is that it is a product of war, occupation and violence, and that, over time, displacement has recursively induced conflict. In particular, the violence and destruction of the 1999 post-referendum vote for national independence, together with the most recent social unrest in Dili in 2006, have drawn attention to conflict-induced displacement over development-induced displacement. This perception of conflict has, in turn, shaped government policy on land and how land relations are understood. I highlight the historical processes of state territorialisation and development interventions, to emphasise that they have produced as much displacement as conflict, and that conflict-induced displacement and development-induced displacement are not mutually exclusive but implicate one another (Muggah 2003).

The modern state seeks to exercise and consolidate its authority within its claimed geographical boundaries through various practices and processes that enable it to control the populace and natural resources; and these strategies can be termed 'territoriality' and 'internal territorialisation' (Vandergeest & Peluso 1995). 'Territorialisation' processes include the establishment of formal regulations on land boundaries, property rights, political administration, and population move-ment. State 'development' interventions can be included in the modern state's portfolio of extending its spatial and administrative powers (Ferguson 1994; Vandergeest 2003). Individuals and other politico-legal institutions (e.g. customary authority figures) may also engage in their own practices of territorialisation to assert control over land and natural resources, which can come into conflict with state forms of territorialisation (Peluso 2005).

State territorial control and conflict occurred in tandem during Portuguese and Indonesian rule, resulting in internal displacement. The task of establishing state authority and control over people and land was carried out through the use of military violence to pacify the East Timorese and symbolic violence through redefining administrative boundaries and mapping legal frameworks onto land in order to remake territory. Along with state agricultural land concessions, a series of land titling programs distributed an estimated 2843 Portuguese titles and 44 091 Indonesian titles (Fitzpatrick 2002, p. 202). Most of these titles are in Dili and other urban centres. By the end of the Indonesian occupation, only 3 per cent of the total land area in Timor-Leste was held as 'non-customary primary industry land' pertaining to coffee plantations, timber plantations and transmigration (Nixon 2007, p. 103). The low percentage of state-acquired land suggests that colonial hegemony was incomplete.

State territorialisation strategies under both regimes were strongly influenced by global discourses of the time, including those relating to 'Western civilisation', 'modernisation', and 'development'. For most of the Portuguese era, the colonialists only established indirect rule through reliance on allied indigenous chiefs and rulers (liurai) (Gunn 1999, p. 192). However, colonial displacement and dispossession go back to the late nineteenth century when the expansion of colonial military capabilities led to direct and violent encounters with the East Timorese. The rise of colonial capitalism made it necessary to reconfigure the administration of land and labour, allowing the state to exert greater control over territory. Colonial state plantations generated less revenue than smallholder production, indicating that plantations were guided more by political rather than economic considerations, enabling colonial power to expand into the rural interior of the island. These introduced changes generated new forms of conflict and displacement related to taxation, dispossession, and contestation over political legitimacy, and local resistance that resulted in several major anti-colonial revolts. Inter-clan warfare was also a constant component of Timor-Leste's history (Gunn 1999). Arguably, local resistance was due to competing territorialities—between local and state-claimed territorialities. The anti-colonial revolts were met with a violent response by colonial authorities who sought to abolish indigenous kingdoms and chiefdoms and impose a secular local governance structure, new administrative boundaries, and enforced formal regulation of people and land.

After the Second World War the Portuguese administration attempted forced resettlement and sought to develop the agricultural sector in order to bring the East Timorese closer to 'civilised' life. Despite colonial pressure, most East Timorese remained scattered, making it difficult for agricultural extension programs. Commercialised wet rice cultivation was advanced for national development, and had a strong impact on local migration patterns as people moved to work on rice fields (Metzner 1977). With the rise of nationalist movements in the African colonies, and a poorly financed colonial administration, Portuguese territorialisation processes came to an abrupt end in 1974. However, the Portuguese-established local governance system was readily adopted by the Indonesian government during its rule, and remains the foundation of the current system.

In a similar fashion, state territorialisation of Timor-Leste flourished under the Indonesian authoritarian government. The pre-invasion population in Timor-Leste was estimated at 688 771 in 1974, and in the invasion years nearly 40 000 people fled over the border into West Timor and a further 4000 fled to Portugal and Australia. By December 1978, 373 000 East Timorese had been forcibly concentrated into an estimated 139–400 'strategic hamlets' (Taylor 1999, p. 90; CAVR 2006, p. 61). Outside these Indonesian-controlled areas were Timorese resistance frontiers. An estimated 102 800–183 000 East Timorese perished and a large proportion were subject to human rights violations from 1974 to 1999 (CAVR 2006, p. 44). After the violent annexation of the territory, some strategic camps were closed down, while other camps, such as the case study discussed below, were transformed into resettlement sites and provided with schools, clinics, markets and transportation. Enforced resettlement was justified to improve local livelihoods by bringing the population closer to new public amenities. A series of major state-driven projects, closely resembling Scott's (1998, p. 4) notion of 'high modernism', were undertaken in the name of bringing welfare to the populations, notably the opening of new roads and bridges to create economic corridors required for socio-economic prosperity. The regime built large-scale agricultural projects and implemented its landmark transmigration program to promote commercialised wet rice cultivation as the primary vehicle of economic development. Many East Timorese were displaced and dispossessed by the conversion of land for large-scale agricultural purposes.

The state ideology of *Pancasila* accompanied technical interventions for 'improvement' and permeated all spheres of local life to instil a sense of 'Indonesian' identity, culture (based on Javanese cultural norms), and attitudes for 'development' (cf. Guinness 1994). Overall, Indonesian territorialisation strategies brought new infrastructure, ideologies and values to the East Timorese.

The aftermath of the 1999 popular consultation vote for national independence, however, took a destructive turn, damaging almost 80 per cent of the physical infrastructure that the Indonesian government had invested in, and displacing over 250 000 East Timorese into West Timor, with the majority in Timor-Leste internally displaced.

To a large degree, state territorialisation practices underlie aspects of internal displacement in Timor-Leste. Violence was implicit and viewed as a necessary part of 'development' by the Portuguese and Indonesian regimes; hence displacement caused by development, territorialisation and conflict is overlaid and interconnected. Clearly, then, conflict-induced displacement and development-induced displacement are less discernible as two distinct phenomena than commonly conceived. This legacy of displacement presents contemporary challenges to the Timor-Leste government.

New state, old trajectory

Taking displacement as a process rather than an event in a specific place and time, the lingering effects of colonial territorialisation took on a recursive dimension in producing displacement and conflict in the post-independence years. In particular, the imbrications of incomplete Portuguese and Indonesian territorialities over local forms of territorial claims and practices have produced multiple and overlapping claims to land and property, leading to fresh tensions and divisions. The 2006 social crisis in Dili marked the worst period of political instability, with nearly 150 000 urban residents internally displaced as generalised violence took on an 'ethnic' dimension. The violence was built on historical and more contemporary political, economic, social, and generational grievances (Kingsbury & Leach 2007).[1]

In contrast to the previous regimes, the Timor-Leste state is taking a less violent approach in establishing its authority. Its management of the internally displaced persons (IDPs) of the 2006 crisis is a case in point. The Timor-Leste state, together with international humanitarian assistance, expeditiously managed the 2006 wave of displacement. After three and a half years, all IDP camps were closed down, with residents either returning to their former places of residence or relocated to new sites. The Timor-Leste government implemented a recovery, resettlement and restitution package (*Hamutuk Hari'i Futuru*), which provided shelter and housing, social protection, economic assistance and security: a stark contrast to the management of displacement in Timor-Leste's past.

Nevertheless, contemporary approaches to territorialisation and 'development' reveal historical continuity. In 2009, the political impasse on attending to contesting land claims ended with the drafting of the Transitional Land Law (2009).[2] Specifically, the proposed articles on 'special adverse possession' and 'community land' are most relevant to situations of displacement and dispossession.[3] The special adverse possession principle enables claimants to obtain land and property title either through long-term and peaceful occupation of land (starting before 31 December 1998), or possession of title based on previous primary or secondary rights under the Portuguese or Indonesian times (Article 21). Compensation will be awarded by the state or successful claimants to unsuccessful claimants (Article 42). The legislation of 'community land' has significant ramifications for both customary landowners and displaced people in the rural districts. The Law defines customary land as 'land in areas where a local community organizes the use of the land and other natural resources by means of norms of a social and cultural nature' (Article 23). Local communities are assumed to perform 'customary norms and practices' as established in Article 25 (1a–c): 'on community land, local communities participate in: the management of natural resources; the resolution of conflicts relating to the use of natural resources; the identification and definition of the boundaries of the lands they occupy'.

The recent national Strategic Development Plan (SDP) for 2011–30 could similarly generate conflict and displacement. Like previous political regimes, the current Timor-Leste government perceives a development 'gap' compared with its 'developed' neighbours, with the dominance of a subsistence economy, high rates of illiteracy and mortality, food insecurity and poverty. Thus development is taken to be 'self-evidently necessary' (Ferguson 1994, p. xiii). The government envisions opening up large agricultural fields, creating regional development corridors and redefining land into 'sustainable agriculture production zones' and 'forest conservation zones' (SDP 2011, pp. 107–16). The Plan also proposes the Millennium Development Goals Suco Program which seeks to build five houses per hamlet in every village across the country for 'vulnerable people' (SDP 2011, pp. 107–17). Who these vulnerable people are and how land will be acquired by the state to carry out such a project were not, however, laid out in the Plan. These inherently spatial state interventions can potentially directly or indirectly result in more displacement (cf. Vandergeest 2003). The post-conflict development agenda may even exacerbate the negative social consequences of previous government interventions. The national land titling program, which can be taken as another example of state territorialisation, poses potential risks of displacement and dispossession.

Methodology

The historical and current processes of state territorialisation outlined above are discussed for Mulia, an Indonesian-established resettlement site on the north-eastern coast of Laga sub-district (Figure 1). The residents of Mulia were forcibly resettled by the national authorities during the Indonesian occupation. At the time of research, the population in Mulia was 1067, comprised of families originating from four highland villages: Waitame, Gurusa, Afasa, and Baagia. Families from Waitame made up the majority of Mulia's residents, and this group forms the focus of my analysis. It was difficult to estimate the exact population as villagers moved between Mulia and Waitame, while many of the younger generation resided in Dili following the expansion of education and employment opportunities after independence. Fieldwork from 2007–08 employed informal and semi-structured interviews, participatory observation and oral history to investigate the livelihood strategies that rural East Timorese draw on to respond to, and recover from, the immense material, affective, and social impacts of displacement.

Mulia: the 'honourable' Indonesian resettlement village

Upon surrendering to the Indonesian forces in 1978, residents of the highland village of Waitame in Quelicai sub-district were forcibly resettled by the Indonesian military onto the north coast of Laga sub-district. When villagers first arrived at the site, the land was 'unoccupied' and covered with wildly overgrown grass, and they were confined on the camp grounds with populations from Baucau, Baagia, Viqueque, and Los Palos sub-districts. They were forced to clear the land to construct shelters, and survived by scavenging for wild food. Camp residents were closely monitored by the Indonesian military and civilian defence (*hansip*) drawn from the camp population.

Following the pacification period, camp residents were given the choice of returning to their former residence or remaining at the site. Residents originating from the further sub-districts left, but the majority of the residents from Quelicai remained. In the 1980s, Indonesian New Order development transformed the camp into a newly furnished 'model' resettlement village. The camp was officially named Mulia (Indonesian term for 'honourable' or 'supreme'). A primary school, village hall, and church were built with community forced labour. Government housing was also constructed for migrant Indonesian workers, such as government functionaries. By the early 1990s, the shift in national focus from territorial security to economic development began to have significant impacts on local livelihoods. Mulia was connected to the electricity grid, public transport became available and access to markets was enabled. There was increased freedom of movement, and the newly opened road to Quelicai enabled families to move between their ancestral settlements and Mulia to make the most of available resources.

Competing land claims

Since the customary landowners of Mulia surrendered later than the settlers from Quelicai, their land was already occupied when they returned. In Portuguese times, most customary landowners resided in the present-day village of Tekinomata, even though a handful of families resided in Mulia, particularly those who raised buffaloes and goats and required large tracts of grazing land. Upon their return, customary landowners were dispossessed of their residential and grazing land. Nevertheless, a small number of customary landowners resided in the newly demarcated Mulia village, but were registered administratively in Tekinomata. Because of the dispossession of land from the landowners, the relationship between landowners and settlers has not improved significantly over three decades. Settlers stated that the Indonesian authorities had compensated the landowners for the expropriated land, and a ritual feast was held to appease the landowners' ancestors. But not all landowners acknowledged these processes. After independence, settlers found themselves no longer residing on 'state land' but instead on contested 'customary land'.

The ongoing land conflict between the residents of Mulia and the landowners of Tekinomata is well known at the national level (see, for example, Meitzner Yoder 2003, p. 11). The everyday tensions between settlers and landowners are interconnected with individual, family, and communal frustrations, such as the lack of economic opportunities, the limited amount of space for grazing livestock (which has resulted in livestock straying into and damaging food gardens) and the extraction of forest materials for firewood. Conflict between the two communities is not always directly concerned with land but is multi-layered, shaped not solely by the local social, cultural, economical, political, and ecological milieu but influenced by wider structural changes. A notable example occurred in the weeks following the 2007 parliamentary election. A youth from Tekinomata, disgruntled over the election results, stoned passing vehicles one evening. The youth injured a passenger of a passing vehicle who was coincidentally from Mulia. Overnight, there were threats made by youths in Mulia to burn the properties and rice fields in Tekinomata. A brawl ensued the following morning in Tekinomata, which was met with police intervention and later resolved through a community-based reconciliation process, but tensions remained.

The persistence of conflict between the two communities is based on several lines of division. First, the settlers had no prior kin and marital relations with the landowners, which greatly limited settlers' ability to assert and negotiate customary land rights to occupy the site. Under Timorese customary tenure systems, access and control of land is reliant on membership of a lineage group or 'house' (*uma kain*), gender, marriage, and social status within customary socio-political domains (Fitzpatrick 2002). 'House' members hold varying degrees of property rights and are able to access family-inherited land and may also enjoy use rights over general areas of land belonging to the group for cultivation. Accordingly, one might expect little adherence to the commands of customary authority in the areas of resettlement, where newly arrived displaced people/settlers have no prior socio-political links to customary landowners (Fitzpatrick 2002). However, inter-group alliances may also translate into land access. Following historical settlement in a particular territory, a local 'order of precedence' may be established on a temporal and spatial basis to differentiate social status, authority and seniority (Fox 1996). The founder-settlers or senior-most origin group usually has authority over land access. Non-kin settlers may be incorporated into the existing local 'order of precedence' through marriage or establishing an alliance with the origin groups. In the absence of a legal land framework, the 'first possession' principle has minimised conflict in rural areas due to its potential to maintain social order but, at the same time, the emerging multiple public authorities have competed for authority, consequently undermining the legitimacy of customary institutions, creating potential for conflict (Fitzpatrick & Barnes 2010, p. 234). Despite three decades of residence in Mulia, there was little inter-marriage between settlers and landowners, and settlers were not incorporated into the local order of precedence.

Second, there are historical political differences between the two communities. Since most of the landowners supported the independence cause, they mistrusted the settlers, some of whom were former members of Timorese militia that supported integration with Indonesia. The landowners described the settlers as occupants (*okupados*), instead of refugees (*refujiados*) or displaced people (*ema dislokado*), which aimed to emphasise that the settlers were not 'victims of war' who were residing on the land out of desperate and dangerous circumstances. Land-owners insisted that the settlers had their own land in Quelicai and that they should leave, since the root causes of displacement had now ceased. Third, the landowners were concerned that their own children would not have sufficient land for shelter and livelihood in the future.

Due to the lack of secure customary tenure rights, settlers turned to the Timor-Leste state to legitimise their land claims on the basis that they were 'victims of war' and coerced by the Indonesian government to resettle, they had occupied the land for nearly 30 years, and they had made significant socio-economic improvements to the land. However, their claim has not been given state recognition. First, the newly introduced formal legal mechanisms for land ownership are limited in applicability to urban and peri-urban areas. Second, attempting to gain formal land rights through the state has been complicated by the contradictory positions held by several sub-district-level government officials who are also customary landowners or members of landowning lineages. These officials have influenced administrative decisions, and in 2003 all administrative responsibilities concerning Mulia were transferred from Laga to Quelicai sub-district. Mulia was reduced to its current status of 'provisional village' (*suku provisório*), and the village name reverted to the customary name of Waiaka. These procedures might have potential political significance for future land claims in Mulia.

Contestation over land ownership does not only occur between settlers and landowners; there are also competing land claims amongst the customary land-owning groups. Mulia is customarily claimed by two lineage groups: Oma Racolo asserted their claim over the site based on mythic-genealogical links, whereas Oma Tameda established their claim based on historical (and ongoing) land use for grazing livestock. According to oral traditions, nearly three generations ago Oma Racolo had given land access to Oma Tameda to raise their livestock. However, it was not clear what degree of customary land rights was negotiated between the groups. The lineage of Oma Tameda is in turn affiliated to the senior lineage of Oma Ina Wai, the ruling political group in colonial times, and whose members currently hold government positions—Oma Ina Wai was thus influential in government decision making, but had no customary claims over Mulia. Contestations over land claims are thus multi-dimensional and layered, and land claims are asserted in a variety of narratives—historical association based on oral history, historical and current land use, and long-term occupation.

Livelihood strategies

A major concern for settlers in Mulia has been the shortage of cultivation land. Mulia is crowded with housing with little arable land. Families had small garden plots (*kintal*) (less than 25 m^2) surrounding their dwellings compared with the average Timorese household with just less than 1 ha of farmland. Previously in Quelicai, families commonly held a *kintal* in addition to fruit trees, and one or two bigger plots of garden, measuring up to 2 ha, further away from dwellings. Because Mulia is located between Tekinomata and Sesal—the customary landholding groups—there was little possibility for physical expansion (see Figure 1). To overcome land shortages most families worked as sharecroppers in the rice fields of the customary landowners. This enduring land access–labour exchange first began under the Indonesian occupation. The authorities recognised that there was little suitable land for cultivation in the resettlement site and authorised landowners to allocate land parcels for settlers who worked in collective farming groups.

Historically, sharecropping arrangements in Tekinomata operated largely amongst members of kin and marital networks. It was also commonplace for the local ruler (*liurai*) to force slaves and commoners to contribute labour in his rice fields. Sharecropping has benefited richer landowning families in Tekinomata, particularly descendants of noble lineages, who owned between two and six parcels (*ulun*) of paddy field. During the occupation, and at present, these landowning families tended to be wage earners due to their privileged upbringing, or raised large herds of cattle that needed constant attention. Consequently, landowners and their children did not devote the time and effort to bring all of their rice fields into production, needing the settlers to meet this labour shortfall through sharecropping. Not surprisingly, the social tension between the two communities

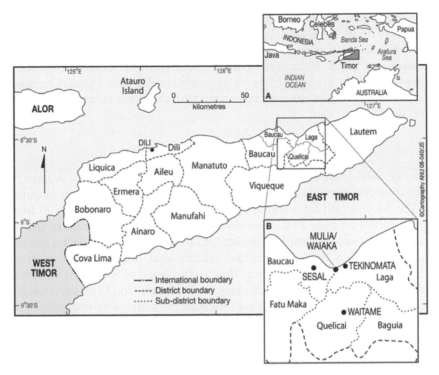

FIGURE 1. Locality map of Timor-Leste and the research.

has seen some settlers denied work as sharecroppers. Many sought work in the neighbouring village of Sesal, where there was less resentment towards them than in Tekinomata, and fields had better irrigation and higher yields.

To overcome the lack of agricultural land, settlers diversified their livelihood strategies to include non-farm activities. Each household typically engaged in a range of agricultural activities including livestock-rearing and fishing. Two road-side food stalls were popular rest stops for passing commuters. Some men sold palm wine by collecting sap from the abundant coastal palm trees. Several women acquired micro-finance to open small kiosks which sold everyday provisions. In the absence of state enforcement of law over natural resources, some settlers quarried rock from the hills behind the settlement or from the Wai'mua riverbed during the dry season. Rocks were broken into small pieces and sold according to size, with smaller rocks commanding higher prices (prices ranged from US$4.50 to $15/m^3). Sales were dependent on occasional construction trucks passing through the village and buying rocks and sand for urban centres. Two timber trading companies in Mulia, established since Indonesian times, imported timber from Indonesian sellers from the neighbouring islands of Alor, Wetar and Sulawesi.

Lack of secured access to land in Mulia has prompted settlers to return to their original village of Waitame to reclaim and re-cultivate family-inherited agricultural land and fruit groves. The interior highland and north coast are in different agro-ecological zones, enabling settlers to migrate seasonally to cultivate large vegetable gardens. Due to the lack of a large permanent population, settlers could acquire customary use rights from the lineage and wider kin group to open up additional garden plots on fallow land. Settlers then migrated back to Mulia during the rice-growing season either to work as sharecroppers or assist relatives in the rice fields by contributing in-kind labour. Thus, there was an economic trade-off between productive food gardens in Waitame and the high-yielding rice fields of Tekinomata and Sesal. Reluctant to move away from the 'modern' government amenities available in Mulia, settlers chose pragmatically both to move between the two sites and make return journeys associated with the socio-cultural significance of ancestral land.

Access to rice fields: sharecropping

The wet rice cultivation season on the north coast is from March to August. From land preparation to harvesting, wet rice cultivation involves the labour of numerous actors. The harvest circulates within the larger community in the form of food and as a ceremonial exchange item. Landowners grow a local rice variety (*fos timor*), believed to taste sweeter than Indonesian and other imported varieties (*fos mutin*). The local variety was highly prized and seldom sold unless there was monetary need. The local variety sells at about double the price of imported varieties, but most families did not travel to larger markets to sell rice due to relatively high rice-milling and transport costs. Household income from rice was normally spent on everyday small goods such as soap, cooking oil, salt, kerosene, sugar and clothing. Larger expenditures such as marriage and mortuary payments and children's school fees were met through monthly wages or sales of livestock.

Two sharecropping arrangements were practised in Tekinomata. Under both arrangements, the landowner provided land access to the paddy field and seeds reserved from the previous harvest for cultivation. Sharecroppers were known simply as 'labourers' (*ema serbisu*) (cf. Metzner 1977, p. 192). In the first arrangement, sharecroppers (mostly working as family units) provided the labour to prepare the paddy for planting, by constructing and maintaining the terraces and irrigation channels. For larger paddy fields, a second type of arrangement engaged sharecroppers who owned buffaloes, or had access to a tractor. In both cases, sharecroppers were responsible for monitoring the crop, weeding, and harvesting. Landowners might contribute labour in each stage of the cultivation. More likely, they would contribute cooked food in exchange for the sharecroppers' labour. Once the rice is harvested, it is threshed and cleaned. During these stages, more people, typically young men, would contribute labour, their labour being remunerated with in-kind payments of a proportion of unhusked rice or the provision of meals. The harvested rice would be divided equally between the landowner and sharecroppers.

An obligatory harvest ritual preceded rice distribution. This ritual, which involved an animal sacrifice, was necessary to appease the ancestors and land spirits in order to restore the field's fertility for the next season. Depending on the landowners, sharecroppers were sometimes obliged to purchase a goat or chicken or contribute a proportion of money for the purchase. The ritual is typically led by the landowner or a ritual expert with historical knowledge of the rice field and well-versed in ritual speech. Despite insistence by both landowners and sharecroppers of their mutual social exclusion, sharecroppers were clearly involved in upholding the moral economy between landowners and the localised spirit realm. By contributing to the purchase of a sacrificial animal, sharecroppers legitimised the landowners' claims over the rice fields. To maintain access to the rice fields, sharecroppers followed certain circumscribed practices, notably being entangled in sustaining this larger moral economy that encompassed the relationships between customary landowners, ancestors and land spirits. Although settlers dismissed customary land claims over Mulia, their participation in sharecropping and the enactment of harvest rituals at the behest of landowners can be seen as vesting customary institutions with authority (cf. Sikor & Lund 2009).

Sharecroppers gained and maintained access to rice paddies in various ways. Competing claims amongst customary landowners advantaged settlers. While rights of access were denied by certain members of the landowning lineages, others readily tapped into settlers' labour for their private interests. Landowners who provided land access stressed that individuals who put in ample labour would get their due reward. Hence, sharecroppers tended to work diligently in order to ensure that the crop did not fail. Access to technology, such as buffaloes and mechanical tractors to plough the paddy, gave another channel of access to settlers. Although landowners might retain control over access to land, and dictate the rules pertaining to labour and resource inputs, however, they could not assert absolute control over production nor gain the ability to derive benefits from land without the labour of the settlers. Sharecropping is thus an enduring mode of livelihood for landowners and settlers alike. The conditions of access, and corresponding power dynamics, between the two groups will nevertheless continue to evolve with the future introduction of formal land laws, market penetration and broader political and economic transformation.

The different social and power relations shaping *access* and *ownership* to customary land are clearly discernible, and both are important in distinct ways in securing shelter and livelihood in Mulia. Relationships between hosts and settlers, limited here to economic cooperation through sharecropping, have meant that tensions still persist over land claims in Mulia. Both landowners and sharecroppers believe that these economic ties have not translated into enduring social ties. If members of either group were faced with social or financial difficulties, they did not seek support from the other, and nor did they engage in shared socio-cultural activities beyond the rice fields. Instead, they relied on their own families and extended kin networks as their socio-economic safety net. Notwithstanding the persisting social divide between the two communities, a 'common ground' existed through sharecropping arrangements, suggesting that customary tenure principles are negotiable and transformative in relation to changes in the political, economic and social milieu.

Implications for rural land access under statutory law

At present, the proposed Transitional Land Law will be applicable only in the urban and peri-urban areas. Nevertheless, the national Strategic Development Plan (2011, p. 112) states that 'an on-request title service will be provided for farmers willing to pay for the service' to facilitate 'progressive farmers' undertaking agricultural development. The Plan highlights a real possibility that land titles may extend to rural districts in the foreseeable future, with formal law having potential impacts on rural populations affected by displacement.

The case of Mulia demonstrates how competing claims to land at the local level are asserted through both actual possession and symbolic actions (cf. Li 1996; Peluso 2005). Unruh (2008, p. 104) contends that national land reform and land policies only succeed by taking account of local land relations that have 'local legitimacy' and 'pervasiveness'. However, the legal system of clarifying land ownership through 'special adverse possession' limits individual claims to either the possession of a previous legal title or long-term physical use and/or occupation of land, both of which dismiss land claims based on origin myths and historical precedence. In addition, there are a multitude of non-title-based claims in rural districts where land is held mainly under customary tenure—which implies that legal solutions alone will be unlikely to clarify competing land claims (Meitzner Yoder 2003), hence there is little evidence of demand for land titling (International Crisis Group 2010). In Mulia, settlers are likely to receive formal land titles under conditions of 'special adverse possession' based on their long-term occupation of land under socio-political duress. Conversely, this legal precept will not work in favour of customary landowners whose claims to land are based on 'origin group authority' rather than continued occupation or use. The undermining of customary authority over land by formal law has the potential to foster fresh conflict between settlers and landowners, which could potentially impact livelihoods on both sides. Considering the persisting tensions between Mulia and Tekinomata, land titles could have the adverse effect of landowners denying settlers access to the rice fields. Since Mulia does not have sufficient arable land, negotiating land access through customary landowners is crucial to support settlers' livelihoods. Furthermore, a number of landowners historically and presently use some land area on Mulia as grazing ground during the dry season, and therefore it is unclear how such land use will be legally demarcated and enforced in practice.

Regulation of customary land under the 'community land' terms of the draft Land Law does not take into consideration the long-term impacts of displacement and dispossession. The definition of East Timorese 'local community' in Article 24 relies heavily on an archetypical conception of rural village life, where genealogically linked groups share a common place of residence and identity, and engage in localised subsistence modes of production. As a long-term impact of displacement, communities may no longer be situated in a single locale, as in Mulia where residents situate themselves between the resettlement site and ancestral land to stake land claims in both sites. The physical use or occupation of customary land at the ancestral settlement may not, however, be a viable option for other displaced communities. Inter-generational effects of displacement may become important. The younger generation of the displaced, those born and raised in the resettlement sites, may have less sense of belonging to the ancestral land, as in Mulia, where youths expressed ambivalent ties to the 'old land'.

The legal articulation of 'customary norms and practices' in the Land Law similarly shows a lack of consideration for the impacts of displacement and dispossession. The definition of 'customary norms and practices' is generalised and assumes that 'East Timorese customs' have remained static despite the evident changes made in local realms under foreign incursions, state formation, market penetration, and, in recent years, foreign aid and development. Such a view ignores the inter-generational impacts of displacement; 'traditional' knowledge and resource management practices may have been lost as a result of protracted displacement. The codification of 'customary norms' risks cementing inherent societal inequalities, such as the marginalisation of women in patriarchal communities.

Considering the complex overlaps between ongoing social conflict and historical land and property disputes, the best way towards effective land management might be a holistic approach through restitution, mediation, and reparation. Any proposed land restitution program will unlikely resolve all land and property disputes in Timor-Leste since the issue is not so much technical in nature, but rather how the country will succeed in transcending the legacy of colonialism and conflict in the long term such that peace, development and reconciliation may be achieved (Du Plessis 2003). Nevertheless, the national symbolic recognition of victims of historical injustices as 'heroes' of national resistance has not been matched by justice in terms of prosecution of war criminals. The government has not followed the recommendations of Timor-Leste's two truth commissions, but has chosen to develop strategic diplomatic relations with Indonesia through reconciliation, thereby forsaking victims' reparation. Overall, the limited under-standing of the lived experiences of displacement is a cause of concern for the formulation of legal measures that seek to protect rights, and may do more harm.

Conclusion

This paper has focused on the legacy of state territorialisation in causing displacement and dispossession, which challenges the overriding narrative of internal displacement in Timor-Leste as a product of conflict. The creation of land titles can be taken as a form of territorialisation that seeks to extend the state's spatial power, and in turn could change local access to land. Drawing on the case study of Mulia, an Indonesian-era resettlement village situated on customarily claimed land, the national focus on clarifying land ownership offers but one dimension to understanding the complex unfolding of displacement 'on the ground'. Although at the community level tensions persist over competing claims to land ownership, settlers have gained access to land for livelihoods by forging economic links at the household level with customary landowners lacking labour to work their rice fields. Land ownership and land access are thus distinct concerns for both settlers and landowners. There were no competing land claims over the rice fields; rather, settlers maintained their access by meeting the prescribed conditions set by landowners.

Analysis of land access is a useful lens to identify and understand the diversity of land tenure and use arrangements in post-conflict settings. It provides a window on the changing conditions tied to land access by drawing attention to why certain individuals and groups are excluded or included (Ribot & Peluso 2003). Conditions of land access do not function according to property relations alone, but are entangled with local historical contingencies, cultural prescriptions, and broader changes in the political and economic context (Ribot & Peluso 2003, p. 157). In Mulia's case, access to land for livelihoods is greatly dependent on labour and capital investment to maintain long-term access. Settlers have shown resilience and innovation to carve out multi-local livelihoods to overcome their plight of displacement and insecure land access. Their insecurity over land access is hindered by the contradictory role of some customary landowners who also hold government positions. Settlers gained access to land through sharecropping primarily as a result of a labour scarcity experienced by landowners, which in turn has been the result of historical circumstances and the current rapid socio-economic transformation that is creating more education and employment in the town centres. Hence, despite control of access to large tracts of rice fields, landowners could not derive benefits from their resource without the labour contribution of settlers.

As a result of socio-political grievances which recursively overlay, combine and transform to create or renew conflict, legal land titles alone are unlikely to resolve the historical tensions over land, such as those between Mulia and Tekinomata. As I have discussed above, land tensions are underscored by socio-political differences. It is beyond the scope of this paper to discuss these interrelated issues. However, displacement and dispossession are not only products of historical conflict; they are also a direct result of a long history of unfinished land titling and development undertakings that have been a source of tension in the past, inform contemporary struggles and may become the basis of future conflicts.

NOTES

[1] The 2006 social crisis began within the national military with a group of disaffected officers known as 'the petitioners'. Violence later erupted between elements of the national military and national police, which caused a breakdown in law and order that was eventually met with international intervention. 'Regional' and colonial-era 'ethnic' identities were politically manipulated, causing entrenched historical divisions to resurface and overlap with more contemporary grievances.

[2] The draft Land Law is intended to apply for an interim period to resolve contesting land ownership. In March 2012, the then President of Timor-Leste Jose Ramos Horta rejected three land laws on the grounds that they will not benefit all Timorese citizens. The three laws of concern were the Expropriation Law, the Special Regime Law and the Financial Property Fund Law. These laws will be revised and debated again in the National Parliament before promulgation by the President.

[3] The draft Transitional Land Law must be read in relation to the rights enshrined in the Constitution of the Democratic Republic of Timor-Leste. Section 54 of the Constitution stipulates that only East Timorese citizens have the right to private ownership of land. The Constitution lays out additional principles that safeguard land rights, gender equality and citizenship rights. A land claims registration program (*Ita Nia Rai*) is carried out simultaneously to the introduction of the Land Law.

REFERENCES

AGARWAL, B. (2003) 'Gender and land rights revisited: exploring new prospects via the state, family and market', *Journal of Agrarian Change* 3(1), pp. 184–224.

BAINTON, N. (2009) 'Keeping the network out of view: mining, distinctions and exclusion in Melanesia', *Oceania* 79(1), pp. 18–33.

COMMISSION FOR RECEPTION, TRUTH AND RECONCILIATION IN TIMOR-LESTE (CAVR) (2006) *Chega! The CAVR report*, Commission for Reception, Truth and Reconciliation in Timor-Leste, Dili.

CRAMB, R. & SUJANG, P. (2011) '"Shifting ground": renegotiating land rights and rural livelihoods in Sarawak, Malaysia', *Asia Pacific Viewpoint* 52(2), pp. 136–47.

CURRY, G.N. & KOCZBERSKI, G. (2009) 'Finding common ground: relational concepts of land tenure and economy in the oil palm frontier of Papua New Guinea', *Geographical Journal* 175(2), pp. 98–111.

DE SOTO, H. (2000) *The mystery of capital: why capitalism triumphs in the West and fails everywhere else*, Bantam Press, London.

DU PLESSIS, J. (2003) 'Slow start on a long journey: land restitution issues in East Timor, 1999–2001', in Leckie, S. (ed.) *Returning home: housing and property restitution rights of refugees and displaced persons*, Transnational Publishers, Ardsley, NY, pp. 143–64.

FERGUSON, J. (1994) *The anti-politics machine: 'development', depoliticization, and bureaucratic power in Lesotho*, University of Minnesota Press, Minneapolis and London.

FITZPATRICK, D. (2002) *Land claims in East Timor*, Asia Pacific Press, Canberra.

FITZPATRICK, D. & BARNES, S. (2010) 'The relative resilience of property: first possession and order without law in East Timor', *Law and Society Review* 44, pp. 205–38.

FOX, J.J. (1996) 'The transformation of progenitor lines of origin: patterns of precedence in eastern Indonesia', in Fox, J.J. & Sather, C. (eds) *Origins, ancestry and alliance: explorations in Austronesian ethnography*, Research School of Pacific and Asian Studies, Australian National University, Canberra, pp. 130–53.

GUINNESS, P. (1994) 'Local Society and Culture', in Hill, H. (ed.) *Indonesia's New Order: the dynamics of socio-economic transformation*, Allen & Unwin, St Leonards, NSW, pp. 267–304.

GUNN, G.C. (1999) *Timor Loro Sae: 500 years*, Livros do Oriente, Macau.

GUNTER, J. (2007) 'Communal conflict in Viqueque and the "charged" history of '59', *Asia Pacific Journal of Anthropology* 8(1), pp. 27–41.

HALL, D., HIRSCH, P. & LI, T.M. (2011) 'Licensed exclusion: land titling, reform and allocation', in *Powers of exclusion: land dilemmas in Southeast Asia*, University of Hawai'i Press, Honolulu, pp. 27–59.

HARRINGTON, A. (2007) 'Ethnicity, violence and property disputes in Timor-Leste', *East Timor Law Journal*, available from: http://www.eastimorlawjournal.org/ARTICLES/2007/ethnicity_violence_land_property_disputes_timor_leste_harrington.html (accessed 20 January 2011).

HUGHES, H. (2004) 'The Pacific is viable!', *Issue Analysis* No. 53.

INTERNATIONAL CRISIS GROUP ICG (2010) *Managing land conflict in Timor-Leste*, Asia Briefing No. 110, Brussels/Dili, available from: http://www.crisisgroup.org/en/regions/asia/south-east-asia/timor-leste/B110-managing-land-conflict-in-timor-leste.aspx (accessed 12 September 2010).

ITA NIA RAI (2005) *Strengthening property rights in Timor-Leste Fact Sheet*, USAID Timor-Leste, available from: http://www.itaniarai.tl/eng/about.html (accessed 5 January 2011).

KINGSBURY, D. & LEACH, M. (2007) 'Introduction', in Kingsbury, D. & Leach, M. (eds) *East Timor: beyond independence*, Monash University Press, Clayton, pp. 1–18.

KOCZBERSKI, G., CURRY, G. & IMBUN, B. (2009) 'Property rights for social inclusion: migrant strategies for securing land and livelihoods in Papua New Guinea', *Asia Pacific Viewpoint* 50(1), pp. 29–42.

LI, T.M. (1996) 'Images of community: discourse and strategy in property relations', *Development and Change* 27(3), pp. 501–27.

LUND, C. (2000) *African land tenure: questioning basic assumptions*, Drylands Programme, International Institute for Environment and Development, London.

MANJI, A. (2006) *The politics of land reform in Africa: from communal tenure to free markets*, Zed Books, London and New York.

MEINZEN-DICK, R.S. & MWANGI, E. (2009) 'Cutting the web of interests: pitfalls of formalizing property rights', *Land Policy* 26(1), pp. 36–43.

MEITZNER YODER, L.S. (2003) 'Custom and conflict: the uses and limitations of traditional systems in addressing rural land disputes in East Timor', regional workshop on Land Policy and Administration for Pro-Poor Rural Growth, Dili.

METZNER, J.K. (1977) *Man and environment in Eastern Timor: a geoecological analysis of the Bacau-Viqueque area as a possible basis for regional planning*, Australian National University, Canberra.

MUGGAH, R. (2003) 'A tale of two solitudes: comparing conflict and development-induced internal displacement and involuntary resettlement', *International Migration* 41(5), pp. 5–31.

NIXON, R. (2007) 'Challenges for managing state agricultural land and promoting post-subsistence primary industry development in East Timor', in Shoesmith, D.R. (ed.) *The *crisis in Timor-Leste: understanding the past, imagining the future*, Charles Darwin University Press, Darwin, pp. 101–15.

PELUSO, N.L. (2005) 'Seeing property in land use: local territorializations in West Kalimantan, Indonesia', *Danish Journal of Geography* 105(1), pp. 1–15.

PELUSO, N.L. & LUND, C. (2011) 'New frontiers of land control: introduction', *Journal of Peasant Studies* 38(4), pp. 667–81.

RIBOT, J.C. & PELUSO, N.L. (2003) 'A theory of access', *Rural Sociology* 68(2), pp. 153–81.

SCAMBARY, J. (2009) 'Anatomy of a conflict: the 2006–2007 communal violence in East Timor', *Conflict, Security and Development* 9(2), pp. 265–88.

SCOTT, J.C. (1998) *Seeing like a state: how certain schemes to improve the human condition have failed*, Yale University Press, New Haven and London.

SIKOR, T. & LUND, C. (2009) 'Access and property: a question of power and authority', *Development and Change* 40(1), pp. 1–22.

STRATEGIC DEVELOPMENT PLAN (SDP) (2011) *Timor-Leste Strategic Development Plan 2011–2030*, Government of Timor-Leste, Dili.

TAYLOR, J.G. (1999) *East Timor: the price of freedom*, Zed Books, London and New York.

TRANSITIONAL LAND LAW (2009) *Special regime for the determination of ownership of immovable property*, draft version for public consultation, Ministry of Justice, Dili.

UNRUH, J.D. (2008) 'Towards sustainable livelihoods after war: reconstituting rural land tenure systems', *Natural Resources Forum* 32(2), pp. 103–15.

VANDERGEEST, P. (2003) 'Land to some tillers: development-induced displacement in Laos', *International Social Science Journal* 55(1), pp. 47–56.

VANDERGEEST, P. & PELUSO, N.L. (1995) 'Territorialization and state power in Thailand', *Theory and Society: Renewal and Critique in Social Theory* 24(3), pp. 385–426.

Index

INDEX